Praise for *Statistical Techniques for Forensic Accounting*

"Financial statement fraud has never been a more serious threat to the integrity of our capital markets than it is today. Professor Dutta discusses the auditing and statistical tools available to detect and investigate financial fraud, set against a colorful backdrop of fraudsters, whistleblowers, and corporate scams that proves fact is stranger than fiction."

—**Dennis Caplan**, University at Albany, New York

"Quantified evidence is often most persuasive and is often required to propose adjustments or prove damages. All parties (management, auditors, regulators, litigants) are best served by applying defensible, well-thought-out techniques when estimating proportions or amounts. Professor Dutta's book uniquely addresses a variety of techniques that could be or are currently applied in supporting arguments and determining amounts. A number of these are illustrated with audit application examples."

—**Lynford Graham**, CPA, PhD, CFE, Bentley University, Massachusetts

"Financial crime causes severe damages to capital markets. It not only affects investors who were deceived, but also reduces overall market return through dissipation of trust after fraud scandals. This causes a demand to detect and prevent financial crime in a timely manner. However, financial crime has been deliberately perpetrated by financial or accounting experts, which could not be detected by simple analytical techniques. Advance techniques such as sophisticated statistical methods are more useful in detecting financial fraud schemes. This book will benefit researchers who conduct research using these techniques to detect financial statement fraud. Other parties, such as auditors and regulators, could use them in developing financial statement fraud detection models."

—**Thawatchai Kiatkwankul**, Security and Exchange Commission, Thailand

"I am very pleased to see this book by Prof. Saurav Dutta. It provides a rich discussion on statistical concepts within the context of forensic accounting and fraud detection. It covers topics from why and how fraud is committed, to how one can detect it by using statistical techniques. He has used simple, familiar examples to illustrate the statistical concepts applicable to forensic accounting and fraud. I wish him great success."

—**Rajendra P. Srivastava**, Ernst and Young Professor, University of Kansas

"Professor Saurav has advanced the accounting profession by providing students of forensic accounting with a resource that combines a superb overview of accounting tools for fraud prevention with a careful introduction to the data-mining and statistical tools needed for fraud detection."

—**Glenn Shafer**, Dean, Rutgers Business School, Rutgers University, New Jersey

"In a passage cited by Professor Dutta, Holmes comments to Watson that: 'It is a capital mistake to theorize before one has data. Insensibly one begins to twist facts to suit theories, instead of theories to suit facts.' Professor Dutta's treatise demonstrates not only the methods used to collect, assemble, and classify data, but, far more importantly, how to transform those data into evidence—in short, how to develop theories to suit the facts available. The essential step in this analysis is, of course, inferential statistics. This path leading from data to evidence is expertly navigated by Professor Dutta in terms that will prove understandable and useful to accountants, auditors, and legal professionals engaged in forensic accounting."

—**Dave Marcinko**, Skidmore College, New York

STATISTICAL TECHNIQUES
FOR FORENSIC
ACCOUNTING

STATISTICAL TECHNIQUES FOR FORENSIC ACCOUNTING

UNDERSTANDING THE THEORY AND APPLICATION OF DATA ANALYSIS

Saurav K. Dutta

Vice President, Publisher: Tim Moore
Associate Publisher and Director of Marketing: Amy Neidlinger
Executive Editor: Jeanne Glasser
Operations Specialist: Jodi Kemper
Marketing Manager: Lisa Loftus
Cover Designer: Alan Clements
Managing Editor: Kristy Hart
Project Editor: Elaine Wiley
Copy Editor: Chrissy White
Proofreader: Debbie Williams
Indexer: Christine Karpeles
Compositor: Nonie Ratcliff
Manufacturing Buyer: Dan Uhrig

© 2013 by Saurav K. Dutta
Publishing as FT Press
Upper Saddle River, New Jersey 07458

This book is sold with the understanding that neither the author nor the publisher is engaged in rendering legal, accounting, or other professional services or advice by publishing this book. Each individual situation is unique. Thus, if legal or financial advice or other expert assistance is required in a specific situation, the services of a competent professional should be sought to ensure that the situation has been evaluated carefully and appropriately. The author and the publisher disclaim any liability, loss, or risk resulting directly or indirectly, from the use or application of any of the contents of this book.

FT Press offers excellent discounts on this book when ordered in quantity for bulk purchases or special sales. For more information, please contact U.S. Corporate and Government Sales, 1-800-382-3419, corpsales@pearsontechgroup.com. For sales outside the U.S., please contact International Sales at international@pearsoned.com.

Company and product names mentioned herein are the trademarks or registered trademarks of their respective owners.

All rights reserved. No part of this book may be reproduced, in any form or by any means, without permission in writing from the publisher.

Printed in the United States of America

First Printing June 2013

ISBN-10: 0-13-313381-8
ISBN-13: 978-0-13-313381-3

Pearson Education LTD.
Pearson Education Australia PTY, Limited.
Pearson Education Singapore, Pte. Ltd.
Pearson Education Asia, Ltd.
Pearson Education Canada, Ltd.
Pearson Educación de Mexico, S.A. de C.V.
Pearson Education—Japan
Pearson Education Malaysia, Pte. Ltd.

Library of Congress Cataloging-in-Publication Data is on file.

To My Sons
Saahil and Samir
The Personification of my Aspirations

Table of Contents

Foreword xiii
Acknowledgments xv
Preface xviii

1   Introduction: The Challenges in Forensic Accounting 1
    1.1   Introduction 1
    1.2   Characteristics and Types of Fraud 3
    1.3   Management Fraud Schemes 7
    1.4   Employee Fraud Schemes 11
    1.5   Cyber-crime 17
    1.6   Chapter Summary 18
    1.7   Endnotes 19

2   Legislation, Regulation, and Guidance Impacting Forensic Accounting 21
    2.1   Introduction 21
    2.2   U.S. Legislative Response to Fraudulent Financial Reporting 22
    2.3   The Emphasis on Prosecution of Fraud at the
          Department of Justice 24
    2.4   The Role of the FBI in Detecting Corporate Fraud 26
    2.5   Professional Guidance in SAS 99 27
    2.6   Chapter Summary 28
    2.7   Endnotes 29

3   Preventive Measures: Corporate Governance and Internal Controls 31
    3.1   Introduction 31
    3.2   Corporate Governance Issues in Developed Economies 33
    3.3   Emerging Economies and Their Unique Corporate
          Governance Issues 34
    3.4   Organizational Controls 39
    3.5   A System of Internal Controls 41
    3.6   The COSO Framework on Internal Controls 46

	3.7	Benefits, Costs, and Limitations of Internal Controls 52
	3.8	Incorporation of Fraud Risk in the Design of Internal Controls 56
	3.9	Legislation on Internal Controls 58
	3.10	Chapter Summary 58
	3.11	Endnotes 60
4	**Detection of Fraud: Shared Responsibility** **61**	
	4.1	Introduction 61
	4.2	Expectations Gap in the Accounting Profession 64
	4.3	Responsibility of the External Auditor 66
	4.4	Responsibility of the Board of Directors 68
	4.5	Role of the Audit Committee 71
	4.6	Management's Role and Responsibilities in the Financial Reporting Process 75
	4.7	The Role of the Internal Auditor 78
	4.8	Who Blows the Whistle 80
	4.9	Chapter Summary 84
	4.10	Endnotes 85
5	**Data Mining** **89**	
	5.1	Introduction 89
	5.2	Data Classification 91
	5.3	Association Analysis 93
	5.4	Cluster Analysis 95
	5.5	Outlier Analysis 98
	5.6	Data Mining to Detect Money Laundering 100
	5.7	Chapter Summary 103
	5.8	Endnotes 103
6	**Transitioning to Evidence** **105**	
	6.1	Introduction 105
	6.2	Probability Concepts and Terminology 106
	6.3	Schematic Representation of Evidence 108
	6.4	Information and Evidence 110
	6.5	Mathematical Definitions of Prior, Conditional, and Posterior Probability 110

	6.6	The Probative Value of Evidence. 114
	6.7	Bayes' Rule 117
	6.8	Chapter Summary. 122
	6.9	Endnote. 123

7 Discrete Probability Distributions 125

	7.1	Introduction. 125
	7.2	Generic Definitions and Notations. 126
	7.3	The Binomial Distribution. 127
	7.4	Poisson Probability Distribution. 135
	7.5	Hypergeometric Distribution 140
	7.6	Chapter Summary. 145
	7.7	Endnotes. 147

8 Continuous Probability Distributions. 149

	8.1	Introduction. 149
	8.2	Conceptual Development of Probability Framework. 150
	8.3	Uniform Probability Distribution. 156
	8.4	Normal Probability Distribution. 157
	8.5	Testing for Normality 168
	8.6	Chebycheff's Inequality 170
	8.7	Binomial Distribution Expressed as a Normal Distribution. 171
	8.8	The Exponential Distribution 172
	8.9	Joint Distribution of Continuous Random Variables. 173
	8.10	Chapter Summary. 176

9 Sampling Theory and Techniques 179

	9.1	Introduction. 179
	9.2	Motivation for Sampling 180
	9.3	Theory Behind Sampling. 181
	9.4	Statistical Sampling Techniques 182
	9.5	Nonstatistical Sampling Techniques. 186
	9.6	Sampling Approaches in Auditing 189
	9.7	Chapter Summary. 191
	9.8	Endnotes. 193

10 Statistical Inference from Sample Information ... 195
- 10.1 Introduction ... 195
- 10.2 The Ability to Generalize Sample Data to Population Parameters ... 196
- 10.3 Central Limit Theorem and non-Normal Distributions ... 199
- 10.4 Estimation of Population Parameter ... 200
- 10.5 Confidence Intervals ... 203
- 10.6 Confidence Interval for a Large Sample When Population Standard Deviation Is Known ... 205
- 10.7 Confidence Interval for a Large Sample When Population Standard Deviation Is Unknown ... 209
- 10.8 Confidence Intervals for Small Samples ... 211
- 10.9 Confidence Intervals for Proportions ... 213
- 10.10 Chapter Summary ... 214
- 10.11 Endnote ... 218

11 Determining Sample Size ... 219
- 11.1 Introduction ... 219
- 11.2 Computing Sample Size When Population Deviation Is Known ... 220
- 11.3 Sample Size Estimation when Population Deviation Is Unknown ... 222
- 11.4 Sample Size Estimation for Proportions ... 225
- 11.5 Chapter Summary ... 228

12 Regression and Correlation ... 231
- 12.1 Introduction ... 231
- 12.2 Probabilistic Linear Models ... 232
- 12.3 Correlation ... 233
- 12.4 Least Squares Regression ... 234
- 12.5 Coefficient of Determination ... 236
- 12.6 Test of Significance and p-Values ... 237
- 12.7 Prediction Using Regression ... 238
- 12.8 Caveats and Limitations of Regression Models ... 239
- 12.9 Other Regression Models ... 242
- 12.10 Chapter Summary ... 245

Index ... 249

Foreword

The Supreme Court's 1993 *Daubert v. Merrell Dow Pharmaceuticals* decision has had far reaching consequences on the Federal Rules of Evidence and has set the standard for admitting expert testimony in federal courts. Daubert confirmed that statistics was a field of scientific knowledge and generally admissible under Rule 702 of the Federal Rules of Evidence.

If statistics can be used in cases involving scientific and medical research as in Daubert, and in opinion surveys, market analyses, and even in determining who will be the President of the United States, statistical evidence can also be used by forensic accounting experts in support of their findings.

Statistical evidence must be *precise* and *reliable* to be able to pass Daubert challenges and, if all such challenges are passed, be persuasively used in court to present the expert's opinion and to withstand opposing expert rebuttal and cross-examination by opposing counsel. Precision and reliability are words that also have very important statistical definitions.

Taking a "sample" from a population and drawing a conclusion about some characteristic of that population is something accountants do all the time; however, this is usually done in the context of an audit of financial statements, where data is already accumulated in journals and summarized into ledgers, and the accountant/auditor has had an opportunity to study the internal control environment. Moreover, financial statement audit conclusions are based on the results of many interrelated tests, such as confirmation with third parties, analytical analysis, and timely physical inspections, not just the result of the audit sample. This is often not the case in a forensic investigation.

Unfortunately, when accountants/auditors are taken out of the financial statement audit comfort zone and are engaged to conduct a forensic investigation, they often apply the audit sampling approach permitted by their professional rule-making body that permits reliability and precision to be expressed qualitatively instead of quantitatively. This can only lead them into trouble. It is the ability to quantitatively measure reliability and precision that makes statistical sampling the only real choice in litigation support services engagements.

EisnerAmper was engaged by Counsel ten years ago to provide assurances to the Court regarding the accuracy of 950,000 claims processed in the $6.1 billion WorldCom Securities Litigation Settlement Fund. A random sample across the population of claims that had over $46 billion in recognized loss amounts where over 95% of the claims were under $5,000 would yield a sample of a high proportion of small claims and a small proportion of large claims. Such a test would bias the results against finding the true error in the population of claims. I knew I had to reach out to someone who understood these ramifications and could communicate the approach taken to address them, and the results and conclusions reached, in a manner that would withstand scrutiny from many constituents on the one hand, but be communicated in a manner that would be understood by the Court, so it could be confident in authorizing payments to claimants. I reached out to the only person I knew who could do this, Dr. Saurav Dutta.

EisnerAmper has consulted with Dr. Dutta many times in past ten years in other litigation support services engagements and forensic investigations requiring the use of sophisticated statistical sampling applications and complex assignments involving derivatives, and he has proved to be a valuable resource to our firm.

It was natural progression for Dr. Dutta to apply his technical and communication skills to the writing of a book for forensic accountants and investigators, which can be used by attorneys and triers of fact as well. I have had the pleasure of reading this book during its development and believe you will find the sometimes complex word of statistics explained in a clear and concise manner, and its application will enhance your practice.

David A. Cace
Partner
EisnerAmper LLP

Acknowledgments

In March of 2011, when I was with the Securities and Exchange Commission, Ms. Jeanne Glasser, an Executive Editor at FT Press, contacted me to discuss publishing opportunities. SEC regulations precluded me from working on the book at the time. Exactly a year later, Jeanne reminded me that the publishing opportunity was still available and asked me to start writing. In sum, the book would not be possible without Jeanne's perseverance. Throughout the process, she was always available and provided necessary help and guidance. Moreover, she was patient during the inevitable delays and missed deadlines.

I was introduced to the topics of statistics and accounting during the course of my graduate study at the University of Kansas, for which I am forever indebted to the support of the faculty at KU. In particular, I am grateful to Professors Glenn Shafer and Rajendra P. Srivastava, without whose guidance and mentoring I would not possess the necessary skills to write this book. All of my coursework in probability and statistics were with Glenn. He also gave me my first copy of Strunk and White's *Elements of Style*. Raj relentlessly encouraged me to pursue graduate studies in accounting and chaired my doctoral dissertation.

I benefitted much from my interactions with many of my faculty colleagues at Rutgers University and at the University at Albany. My interactions with Dr. Lynford E. Graham were extremely valuable. Additionally, I thank many organizations for providing me with opportunities to apply statistical techniques to real-world accounting problems. Specifically, I thank Mr. Nicholas Adelizzi, Mr. Victor Albanese, Mr. David Cace, Ms. Robin Cramer, Mr. Tobias Feinerman, Ms. Marisol Gonzalez, Mr. Elliott Lee, Mr. Nicholas Sheridan, Ms. Dayna Shillet, and Mr. Timothy Van Noy for their interactions and practical insights.

The book has benefitted significantly from comments and suggestions of two of my esteemed colleagues, Professors Dennis Caplan and David Marcinko. Both painstakingly reviewed earlier drafts of the chapters and made numerous suggestions to improve the content and the delivery. I am indebted to them for their time and generous help throughout the process. Scott Cestone, one of my former students, tirelessly worked in editing the chapters and providing a student perspective to the material. He did so while simultaneously preparing for his CPA examination, for which I am extremely grateful. I thank Ms. Elaine Wiley and Ms. Chrissy White of FT Press for their editorial suggestions and help in getting this manuscript to its published form.

About the Author

Dr. Saurav K. Dutta is an Associate Professor at the Department of Accounting, Business Law, and Taxation at the State University of New York at Albany, where he previously served as the Chairman of the Department. He has taught at the Graduate School of Management, Rutgers University, and at Zicklin College of Business, Baruch College, New York. He holds a Bachelor of Technology Degree in Aerospace Engineering from the Indian Institute of Technology (Bombay) and a Ph. D. in Accounting from the University of Kansas.

His research interests are in applying statistical and analytic methodology to accounting and auditing problems, and his current work involves analyzing problems in financial reporting, as well as studying the accounting aspects of corporate sustainability initiatives. He has published over 25 research papers in academic journals, including *Auditing: A Journal of Practice and Theory; Journal of Accounting, Auditing, and Finance; Journal of Accounting and Public Policy; Issues in Accounting Education; Journal of Cost Management; Journal of Corporate Accounting and Finance; International Journal of Technology Management; The Quality Management Journal; Corporate Social Responsibility and Environmental Management; Strategic Finance;* and others. Dr. Dutta has presented his research findings at numerous national and international academic conferences and has conducted research seminars at many universities including, Harvard, Oxford, New York University, Rutgers, University of Southern California, Michigan State, Bentley, and Maastricht. He has conducted professional teaching and training seminars at Dai-Ichi-Kangyo Bank, Merrill Lynch, Prudential Insurance Company, and KPMG LLP. He has been invited by the AICPA to conduct workshops on the use of statistics in forensic accounting, and he has also been the "Featured Speaker" for the Corporate Director's Group.

Dr. Dutta has been engaged to design and analyze statistical tests on numerous accounting/litigation projects under the jurisdiction of the New York State Attorney General's Office, U.S. District Court of the Southern District of New York, and the Securities and Exchange Commission, among others. Some of these engagements involved designing statistical procedures to verify claims for settlements of amounts ranging from $500 million to $6.1 billion and include the settlements for MCI-WorldCom, Global Crossing, Cendant Corp, and HealthSouth. He was involved with the reparations of more than 400 million

CHF from the Swiss banks, under the purview of the U.S. District Court of Eastern New York. He has also been engaged to evaluate accounting systems related to hedge accounting, fair value accounting, and mergers and acquisition. Since 2006 he has served as the Subject Matter Expert (SME) for the IMA in their preparation and updating of the CMA examination study guide.

Preface

Unfortunately forensic accounting is a growing profession. In fact, the FBI has added staff to its Forensic Accounting and Financial Crimes Division to address the increase in financial crimes. Further, forensic accounting is an expanding practice for the Big Four and other public accounting firms. This growth is attributable to the increased cost of fraud, the rise of Internet technologies that facilitate fraud, and recent legislation in some jurisdictions. Financial fraud is a global problem, and recent surveys conducted by KPMG and PwC show an increase in both the number and magnitude of fraud in many countries. Globalization, technology, and pursuit of business efficiency have made fraud easier to commit and harder to detect, making the forensic accountant's job more challenging. The ability to detect financial fraud, honed by years of experience, is an art. It is the art of skillful conjecture.

Many useful books are available to help forensic accountants develop this art of conjecture. These pioneering books laid the foundation for the discipline. Donald Cressey's *Fraud Triangle*, Steve Albrecht's books on fraud examination methodology and Joseph Wells' many books outlining fraud schemes are important contributions to the discipline. More recently, Mark Nigrini's book, *Forensic Analytics* applies Benford's Law to develop a tool for fraud detection.

This book, *Statistical Techniques for Forensic Accounting,* builds on the foundation of probability and statistical theory to help readers apply mathematical tools to the art of identifying financial fraud in that it provides a structure to conjecture. Financial information typically consists of large amounts of data. An occurrence of fraud or misrepresentation creates a pattern of errors within the data. Effective forensic accountants uncover those patterns based on intuition, conjecture, or experience. After the pattern has been identified, the onus rests with the forensic accountant to convince others—and ultimately the justice system—that the pattern was deliberate and not caused by random occurrence. This book makes a contribution to formalizing the process of pattern recognition and establishing the probability that it is deliberate. It requires no prior knowledge in probability and statistics and presents mathematical concepts, notation, and equations in a manner understandable to practicing accountants.

Statistics provides a structured process to synthesize and analyze large amounts of data as well as examines uncertainty. Additionally, the discipline has developed sophisticated and well-accepted tools that enhance the efficiency and effectiveness of the process. It is therefore natural that statistical tools such as data mining can be applied to financial data analysis to investigate the possibility of fraud. Statistical methodology has been gaining wider applicability and acceptance in the judicial system in general. For example, DNA evidence and ballistics are built on a foundation of probabilistic reasoning. Anecdotally, jurors often place more credence on DNA evidence, which is probabilistic, than on eye-witness accounts, which are often and notoriously unreliable.

Prior to the use of statistical evidence in forensic accounting, an understanding of the applicability and limitations of such methods is imperative. It is a widely accepted view that mathematics is often best learned within the context of the discipline itself. However for the discipline of forensic accounting, no such book is currently available. It is therefore the primary aim of this book to fill that gap. Several real-world cases of financial various parts of the world are briefly summarized and form a backdrop for the statistical methodology.

1

Introduction: The Challenges in Forensic Accounting

> *I spent about ten years exposing corporate and financial fraud for Barron's magazine, and I found a lot to write about.*
> —Ben Stein

1.1 Introduction

Investigating corporate fraud cases is one of the highest priorities of the United States Federal Bureau of Investigation (FBI). At the end of 2011, 726 fraud cases were being actively investigated by various FBI field offices throughout the U.S.,[1] which was an approximately 10% increase from the number of cases being pursued at the end of 2010 and a 37% increase over a five-year period. It is estimated that several of these cases resulted in losses to investors exceeding $1 billion.

In response to the growing importance of accounting investigative skills within the Agency, the FBI established a Forensic Accountant Unit in March 2009 to support FBI investigations requiring financial forensics. The creation of this program was the culmination of the FBI's efforts to advance and professionalize its financial investigative capabilities. The mission of this unit is to develop, manage, and enhance the FBI's Forensic Accounting and Financial Analysis programs.

Corporate fraud has been a persistent problem in the U.S. economy since the early days of the stock market. One of the oldest and most commonly known frauds is the Ponzi scheme, named after Charles Ponzi, who duped thousands of New England residents into investing in postage stamp speculation back in the 1920s. He promised his investors a 50% return in just 90 days at a time when the annual interest rate at banks was just 5%. He was able to fulfill his promise by using incoming funds to pay off early investors. His name is now synonymous with investment fraud schemes that involve attracting new investor money to fulfill the obligations owed to early investors. As long as the amount of incoming funds in the form of investment by new or existing investors exceeds the outgoing funds that are withdrawn by outgoing investors, the scheme thrives. However, it quickly unfolds in an economic downturn when investors have little excess funds to invest and more investors wish to cash out their investments. In the recent financial crisis of 2007, many investment funds, including that of Bernie Madoff and Stanford International, were found to have engaged in the decade-old Ponzi scheme to defraud investors. In the 1920s, Charles Ponzi's investors lost upwards of $20 million. About a century later, in 2007 Bernie Madoff's investors lost upwards of $20 billion.[2] Although there are obvious red flags to detect Ponzi schemes,[3] these are often ignored by trusting investors who are sometimes in awe of the personality, success, and demeanor of the perpetrators.

Investment fund managers who perpetrate Ponzi schemes most often defraud investors by promising high returns. However, other businesses, though not directly engaging in Ponzi schemes, can still defraud investors by painting a false financial picture. By recording fictitious transactions or other accounting gimmicks, a business might inflate income or revenue or understate expenses to paint a rosier picture of its financial prospects. At times, senior officers of public companies with much personal wealth and reputation vested in the financial success of their companies resort to such means. In this chapter you learn about some of the recent well-publicized cases of corporate fraud.

Not all types of corporate fraud are committed by management to misinform investors and other third parties. A majority of occurrences and investigations of fraud involve acts by individuals committed against their organizations. One notable case prosecuted by the FBI in 2011 involved a former vice-president of Citigroup who embezzled more than $22 million from Citigroup over a period of eight years. The perpetrator transferred money from various Citigroup accounts to Citigroup's cash account and then wired the money to his personal bank account at another bank. False accounting entries were created by the perpetrator to conceal the theft. In this chapter you read about common fraudulent schemes undertaken by employees to steal from the organization and the accounting cover-ups undertaken to conceal the crimes.

The remainder of this chapter discusses the nature of fraud schemes, presents the statistics of fraud investigations conducted by the FBI, and examines trends of various types of fraud. It then briefly covers the common management fraud schemes undertaken in recent years. Next is a discussion of common employee fraud schemes, which summarizes the results of a study conducted by the Association of Certified Fraud Examiners (ACFE). In recent years, with the Internet being so widely accessible, coupled with an immense growth in e-commerce, cyber-crime against businesses is rampant. The chapter concludes with an analysis of the ramifications and challenges imposed by this growing crime.

1.2   Characteristics and Types of Fraud

Fraud is not only a theft of assets but also an attempt to conceal it. Misappropriation of assets without an attempt to conceal is merely a theft, which is usually uncovered quickly through normal checks and balances procedures. Concealment distinguishes fraud from theft. As perpetrators attempt to conceal fraud, they might continue to engage in similar misappropriations over an extended period of time. A theft often occurs only once because the victim becomes armed with the knowledge of that theft and takes necessary precautions to deter future occurrences; the victim of a fraud, on the other hand, is usually unaware of the loss, and hence the perpetrator can repeatedly commit the crime. In that sense, fraud is nothing but recurring theft on the similar type of victim by the same perpetrator. In the instance of corporate fraud, there is a sole victim, the organization, and usually there are multiple occurrences of theft that impact that organization.

The FBI characterizes fraud as comprising of deceit, concealment, and or violation of trust. Fraud is not usually dependent on the application or threat of physical force or violence. The FBI has recently investigated and prosecuted many financial crimes, including corporate fraud, securities and commodities fraud, health care fraud, financial institution fraud, mortgage fraud, and others. This section provides an overview of these schemes, and subsequent sections discuss in more detail corporate fraud committed by insiders (employees and management) of an organization.

Figure 1.1 plots the number of cases in each category investigated by the FBI over the seven-year period from 2005 to 2011. The "other" category is comprised of money laundering, insurance fraud, and mass marketing schemes. The plot represents the total number of ongoing investigations at the end of each reporting period. It is important to note that this does not reflect the number of new cases for each year. An upward sloping line indicates that the number of new cases in that category exceeded the number of cases that were settled or resolved. Similarly, a downward sloping line indicates that the number of cases settled for that category was greater than the number of new cases. Thus, a decline in the number of open cases does not necessarily indicate a decrease in that category of crime but could be due to speedier resolution of the pending cases from previous years.

Over the seven-year period, health care fraud cases have been steadily high at about 2,500 cases annually. In contrast, investigations related to mortgage fraud spiked in 2008 and 2009, exceeding the 2,500 mark. This spike was primarily related to investigations following the collapse of the sub-prime lending market in the United States. Investigations related to securities fraud also increased in the 2008–2009 timeframe, but the increase was not as drastic as the increase in mortgage investigations. Open investigations on corporate fraud were relatively steady over the seven-year period. In contrast, investigations of financial institution fraud and others have been declining in recent years. The percentage of growth or decline in the number of cases over the previous year for each category is plotted in Figure 1.2.

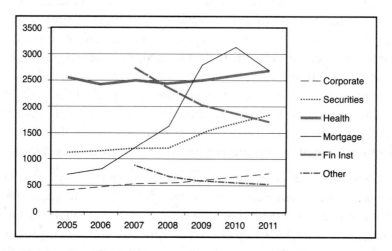

Figure 1.1 Number of fraud cases by category investigated by the FBI between 2005 and 2011

Data for the graphs obtained from the FBI's *Financial Crimes Report to the Public* from the years 2009 and 2011. The 2009 report covers the years 2005–2009, and the 2011 report covers the years 2010–2011.

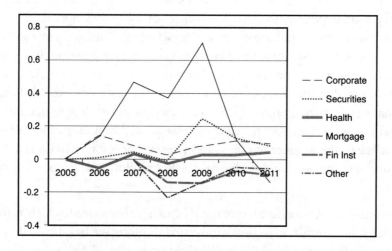

Figure 1.2 Percentage change in cases by category investigated by the FBI 2005–2011

Investigations related to corporate fraud are usually conducted by the FBI in collaboration with the Securities and Exchange Commission (SEC). Often the investigation is initiated at the SEC following a tip from a whistleblower. Upon follow up if the SEC suspects criminal wrongdoing, the FBI is alerted, and a criminal investigation ensues. Corporate fraud cases usually involve accounting schemes designed to deceive investors, auditors, analysts, and others regarding the true financial condition of the corporation. The usual objective for such crimes is to artificially lower the cost of borrowing by inflating the share price based on fictitious corporate performance indicators. As can be seen in Figure 1.2, the number of such cases has been relatively stable over the seven-year period. In 2011, the pursuit of these cases led to 241 convictions, and the FBI secured $2.4 billion in restitution orders.

Corporate fraud investigations focus not only on the misrepresentation of a firm's financial conditions, but also on the investigation of allegations involving insider trading. A multitude of parties can potentially engage in trading based on insider and nonpublic information. These parties include corporate insiders, corporate attorneys, traders, and other financial intermediaries such as accountants and investment bankers who are privy to confidential and nonpublic information.

The volatility of the financial market in recent years has caused investors to seek alternative investment opportunities. This investor need has created a demand for new and innovative investment products and opportunities. Concurrent with investors' growing need for alternate investment vehicles and their eagerness to invest in new and untested products, the FBI saw a steady rise in securities and commodities fraud in 2009 and 2010. These new schemes and trends included

- Securities market manipulation through cyber-intrusion
- Increased commodities fraud
- A continuing rise in Ponzi schemes
- Onslaught of foreign-based reverse merger schemes

Figure 1.2 showed the plot corresponding to securities fraud spikes in 2009, signifying that such fraud schemes are increasing at a faster rate following the recent financial crisis. The victims of securities fraud include individual investors, pension and retirement funds, government entities, financial institutions, and private and public companies. The creation of complex investment vehicles makes prevention and early detection of such schemes difficult, resulting in higher losses for the victims of such schemes.

As seen from Figure 1.1, health care fraud, an important area of investigation for the FBI, has traditionally been their most prolific type of case; however, the number of cases has been relatively stable over the seven-year period if you compare the data to that shown in Figure 1.2: the graph is close to the x-axis, or zero, denoting little change. Data mining techniques, which are covered in a later chapter, are used to detect health care-related fraud schemes. In 2011, the FBI recovered $1.2 billion in restitution, $1 billion in civil settlements, and an additional $1 billion in fines. The prevalent schemes for health care fraud identified by he FBI include

- Billing for services not rendered, either wholly or partially.
- Duplicate billing.
- Upcoding of services to generate higher payments.
- Upcoding of items.
- Kickback schemes.
- Unbundling, which involves billing separately for individual items to maximize reimbursement when they are required to be billed together at a reduced cost. For example, a laboratory test can be ordered individually or as a panel. A panel test usually costs less than the sum of the individual tests.

Another area of fraud that is rapidly increasing is mortgage fraud, the victims of which include financial institutions and investors. Mortgage frauds primarily occur at *entry* or *exit points*. In other words, they occur at the time of loan origination or at the time of foreclosure or delinquency. After underwriting rules were tightened in response to the financial crisis that began in 2007, the year 2011 was the first time that distressed homeowner frauds outnumbered loan origination fraud. As seen in Figures 1.1 and 1.2, the investigations of mortgage fraud spiked considerably in 2009 following the financial crisis and the near-abolition of the sub-prime market.

Following the global financial crisis, many incidents of financial institution fraud came to light. Having started at the highest level in 2007, the cases corresponding to this category have been declining over the years. This includes investigation of financial institution failures. The number of bank failures during 2009 and 2010 had significantly increased to about 150 per year but then in 2011 decreased to under 100 per year. Over the five years following the global financial crisis, there have been about 400 bank failures. Although still a large number, this cumulative total compares favorably to more than 1,000 banks closing over the five-year period from 1987 to 1992 and more than 9,000 bank failures during the Great Depression (1930 to 1933).

Other types of financial fraud investigated by the FBI include insurance scams, money laundering, and mass marketing frauds. One of the most prevalent mass marketing frauds in recent times is the Nigerian email/letter that you might have personally encountered. In this scheme victims are asked to act as U.S. agents to facilitate transfers of huge sums of money held in foreign accounts into the U.S. The victims are promised a generous portion of the total proceeds for their efforts. The victims are then required to open accounts at fraudulent websites and transfer their holdings from legitimate banks to fictitious ones; once completed, the funds are stolen by the perpetrator.

1.3 Management Fraud Schemes

Management fraud, as the name suggests, is perpetrated by the top management of a company who has the intention of misleading investors. The most common form is through accounting manipulation, which materially misstates the financial statements of the company. The motivation behind the fraud is usually to maintain a high stock price and thereby lower the cost of capital for the company.

One of the most common ways to inflate earnings is by simply overstating or misclassifying revenue. There are many well-known fraudulent schemes that have been used to inflate revenue, such as

- Bill and hold sales. In such schemes the company bills the customer for the sale, hence creating a perception that a legitimate sale was made, but it never ships the goods. Instead in the following accounting period it simply reverses the sale, reporting that there was a customer return.
- Booking fictitious sales. This scheme is discussed shortly in the context of fraud committed at HealthSouth.
- Holding books open at the end of the period. Through this method the next period's sales are recorded in the current period, thus inflating the revenue of the current period while understating revenue in the subsequent period. This scheme was used by Computer Associates.
- Delaying reporting of customer returns. When customers return merchandise, the revenue initially recognized from those sales must be reversed. By delaying the reporting of returns to the next period, management effectively reports higher revenue in the current period.

The overstating of revenue would in most cases also overstate accounts receivables and hence would also overstate assets.

The overstatement of revenue may not always be sufficient to inflate income to the desired level; hence management might use a combination of understating expenses along with overstating revenue. The common schemes to understate expenses include

- The understatement of cost of goods sold by padding inventory. When a periodic inventory accounting method is used, the cost of goods sold is indirectly determined by measuring ending inventory. Inventory is overstated, causing a decline in the cost of goods sold expense.

- Capitalization of costs in order to reduce expenses. When costs are capitalized, the resulting expense is spread over multiple years rather than reporting all of it in the current year. Capitalizing costs that should be expensed results in the understatement of expenses in the current period, leading to a overstatement of income. The capitalized costs are reflected as assets, hence also overstating total assets on the balance sheet. This scheme was used in the Worldcom fraud and is discussed in a subsequent section.

- Extending the depreciable lives of assets, thereby reducing depreciation expense. As depreciation expense is linked to the management's estimation of the useful life of an underlying asset, overestimating the lives of assets reduces the depreciation expense, resulting in an overstatement of income. This scheme was employed by Waste Management.

Understating expenses also leads to overstatement of assets or understatement of liability. When costs are capitalized and not expensed, it leads to higher assets. When accrued expenses are not recorded, it leads to an understatement of liability.

Balance sheet fraud is committed with the intention of reporting lower debt and liabilities than the company actually bears. This is often accomplished through the use of off-balance sheet financing. Schemes to fraudulently reduce liabilities include

- Misclassification of leases. Although capital leases are recorded as a liability, under U.S. Generally Accepted Accounting Principles (GAAP), operating leases currently are not required to be reported as a liability. Misclassification of a capital lease as an operating lease allows a company to remove the underlying liability from its balance sheet.

- Not recording accrued expenses. As just discussed, a failure to record accrued expenses results in the income being overstated and the liabilities being understated.

- Concealing liabilities in the accounts of unconsolidated subsidiaries. As in the case of Enron, shell companies were created with the sole purpose of off-loading liabilities from Enron's financials onto the financials of these shell companies.

- Structuring sophisticated financial transactions such as Repo 105 to remove liabilities from the balance sheet. As discussed next, Lehman structured a "round-trip" transaction, which enabled them to reduce both assets and liabilities on the reporting date thereby presenting a lower leverage ratio.[4]

The remainder of this section discusses a few prominent cases of recent business failures in which the use and interpretation of accounting rules were questionable.

Fraud at WorldCom

WorldCom was accused of having inflated profits by $3.8 billion over a period of five quarters. The company undertook the massive fraud by capitalizing costs that should have been expensed. Capitalization of these costs allowed the company to spread the expenses over several years instead of recording all the costs as expense in the current period. Such deferral of costs allowed the company to report lower expenses and therefore inflated income.

From 1998 to 2000 WorldCom reduced reserve accounts held to cover the liabilities of acquired companies resulting in $2.8 billion in additional revenue. They misclassified expenses and marked operating costs as long-term investments. There were undocumented computer expenses of $500 million, which were treated as assets. The fraud was uncovered by the internal auditors in July 2002; soon thereafter WorldCom filed for bankruptcy.

Fraud at HealthSouth

HealthSouth, a publicly traded company headquartered in Birmingham, Alabama, with 1,600 locations spread over all 50 states and three other countries, was by many accounts the first company to be prosecuted under the provisions of the Sarbanes-Oxley Act of 2002. Its former CEO, CFO, and other senior officers fraudulently inflated the company's reported income to meet Wall Street's earnings expectations. The fraud began in mid-1996 and spanned for about seven years, during which time their true cumulative income was $1.7 billion or about 40% of what the company had reported. The SEC charged the senior officers in March 2003 for knowingly falsifying accounting records and designing fictitious entries to overstate cash by $300 million and overstating total assets by at least $800 million.

Evidence presented at the trials showed that facilities owned by HealthSouth submitted legitimate financial reports to the headquarters in Birmingham. However, at the corporate office those numbers were inflated at the time the consolidated financials were prepared. A fictitious account called *contractual adjustments* was created to book fake revenue numbers. Additionally, the company failed to properly record the sale of technology to a related company, resulting in a $29 million overstatement. Also HealthSouth twice recorded a sale of 1.7 million shares of stock in another company, netting a $16 million gain. Examiners also found fictitious assets totaling close to $2 billion.

The fraud mechanism used by HealthSouth required collusion among various employees who were known as "family" and attended quarterly "family meetings" to cook the books. Top company officials reviewed unpublished financial results and compared those with Wall Street expectations. The shortfall was termed as a gap or hole that had to be filled using "dirt." Staff accountants were instructed to make fictitious entries to fill the "gap," and false documents were created in an attempt to conceal the false entries from the auditors. It was common knowledge that auditors verified transactions over $5,000, hence fictitious transactions were made for amounts between $500 and $4,999. To gain a proper perspective on the sheer enormity of the fraud, it required an upward of two million falsified journal entries to overstate the income by almost $5 billion. This was a clear indication of how widespread the involvement and knowledge of the fraud was within the organization. At the conclusion of the investigation, it seemed that everyone in the organization was aware of the massive fraud—except the auditors.

Questionable Accounting Practices at Lehman

In September 2008, Lehman became the largest company in U.S. history to file for bankruptcy. Nine months earlier, Lehman had reported record revenue and earnings for 2007. In early 2008, Lehman's stock was trading in the mid-sixties with a market capitalization of more than $30 billion. Over the next eight months, Lehman's stock lost 95% of its value and was trading around $4 by September 12, 2008. In March 2010, Lehman's Bankruptcy Examiner, Anton Valukas, issued a 2,200-page report that outlined the reasons for the Lehman bankruptcy.[5]

Lehman routinely engaged in short-term borrowing but structured some of these loans under a scheme known as Repo 105. Repo transactions are quite common in the financial industry. Under a Repo agreement a bank borrows funds and transfers assets to the lender as collateral. At a later date, the bank pays back the loan with accrued interest and repossesses the transferred assets, hence Repo. Normally, the collateral is 2% above the borrowed amount. Lehman altered the agreement slightly and transferred assets whose market value was 5% above the borrowed amount. However, instead of classifying these transferred securities as collateral for the loan, to be returned upon the settlement of the loan, Lehman would record the transfer as a sale with an agreement to repurchase on a specified date.

Interestingly, during the term of a Repo 105 transaction, Lehman continued to receive the stream of income through coupon payments from the securities it transferred.

Additionally, just as in an ordinary repo transaction, Lehman was obligated to "repurchase" the transferred securities at a specified date. Moreover, Lehman used the same documentation to execute both Repo 105 and ordinary repo transactions, and these transactions were conducted with the same collateral agreements and substantially with the same counter-parties. Lehman's usage of Repo 105 was timed around the end of reporting periods. The Examiner's Report analyzed the intra-quarter data on the usage of Repo 105 and concluded that its usage spiked at quarter-ends and fell off on an intra-quarter basis. The amount of Repo 105 activity at period-end from late 2007 to mid-2008 ranged from $39 billion to $50 billion. The use of Repo 105 transactions enabled Lehman to remove assets and corresponding debt from its balance sheet, yielding a marked improvement to its leverage ratio. The ratings agencies and counter-parties to Lehman's Repo transactions were concerned about the high leverage ratio of Lehman. Thus, being able to show a decrease in their leverage ratio was beneficial to Lehman. It is unclear as to whether or not the use of Repo 105 led to the demise of Lehman; however, it is evident that Lehman employed a questionable accounting treatment of Repo 105 that had no business purpose or economic significance other than to understate their leverage ratio.

Lessons Learned

As illustrated through these examples, management fraud, although not rampant, could have potentially devastating effects on the reputation and sometimes the viability of the company and the auditor. Moreover, management fraud is usually ongoing over several years. Even though cleverly concealed, many employees are aware of the scheme and usually would have raised red flags or otherwise tried to warn auditors and regulators. Complaints from former employees should not be dismissed casually as being vindictive, but due professional care has to be exercised by the auditor. Statistical techniques presented later in the book can perhaps lead to early detection of such fraudulent schemes and limit losses to investors, employees, and society.

1.4 Employee Fraud Schemes

The Association of Certified Fraud Examiners conducted a substantial study to classify occupational fraud cases. The recent edition published in 2010 presented updated descriptive statistics on the occurrence, damages, and so on of occupational fraud. The schemes discussed in this section are the common schemes reported in that study. Further details on the schemes and methods of prevention and detection are available on ACFE's publication.[6] The following sections present a brief description of the common fraud schemes and ways to detect and deter them.

Skimming

Organizations that engage in cash transactions are vulnerable to skimming by their employees. Skimming involves theft of cash, generated usually from sales, prior to its entry into the accounting records. Skimming is a relatively common occurrence in professional practices where fees are collected in cash. The cashier responsible for collection might pocket the cash and not enter the transaction into the accounting records or subsequently delete those records after being entered into the system. Medical practices are particularly vulnerable to this type of fraud, as small amounts of copayments are collected in cash, and the patients are not that particular about obtaining receipts as there isn't an opportunity for a refund. Instituting proper internal control systems that mandate giving receipts to the customers or installing surveillance equipment can mitigate the risk of fraud caused by skimming.

Even for retailers that sell merchandise for cash, theft through skimming is quite prevalent. In some cases, the employees keep the store open beyond regular business hours and pocket the sales made at those times instead of properly recording the sales in the accounting records. For merchandising companies, because a sale involves the exchange of goods, skimming results in inventory shrinkage. That is, there is reduction in inventory totals without corresponding sales. In these instances, a pattern of inventory shrinkage is generated.

Another type of business that is vulnerable to skimming is off-location rental services, in which the on-site manager usually has much autonomy. The manager may rent out property for cash without making the necessary accounting entry or without reporting the rental revenue to the organization. Random on-site inspections and correlating maintenance expenses with rental revenue are two approaches to ensure early detection of such schemes. The statistical technique of correlation, developed in a subsequent chapter, can be applied to identify the pattern between rental revenue and maintenance costs. The application of this procedure enables the identification of locations that are anomalous and perhaps need to be visited.

Lapping or Fraudulent Write-offs

Another common form of skimming is undertaken by mail room employees who are responsible for receiving payments and can therefore skim the checks received. That is, instead of depositing the checks in the company account and logging the payment into the accounting system, the employees would deposit checks into their accounts and steal the funds. This kind of scheme requires a cover-up or falsification of records. Without a cover-up the scheme is unraveled quickly when the company sends a second bill and the customer furnishes a cancelled check as proof of payment. Hence perpetrators of such crimes also need to conceal the theft of the checks. Two common ways to conceal the theft are *lapping* and *fraudulent write-offs*. In a lapping scheme, one customer's payment is posted on another customer's account.[7] For example, assume the perpetrator stole Mr. Smith's payment; next, when Mr. Jones makes his payment, it is applied to Mr. Smith's account, and later when Mr. Wells makes a payment, it is applied to Mr. Jones's account, and so on. Thus, when the perpetrator laps customers' accounts, he repeatedly alters accounting records, even though the theft of funds occurred only at the outset.

The fraudulent write-off of a customer's account is another way to bring the account up-to-date without the organization receiving the payment. As noted earlier, the customer is going to complain after receiving a second bill for the amount he or she has already paid. This can provoke internal investigation of the missing funds. To prevent customers' complaints, the perpetrator of the crime has to keep the second bill from being sent to the customer. The customer will not be billed by the organization if the account is written off. The perpetrator would therefore try to write-off the customer account and steal the funds that the customer sent as payment. This way the funds are stolen, the customer isn't billed repeatedly, and the accounting records balance. Because most large organizations segregate the duties of receiving cash, maintaining accounts receivables records, and authorizing write-offs, collusion between employees has to occur for this scheme to be successful. Chapter 5, "Data Mining," shows how such collusion can be unraveled through the use of association mining.

With the increased popularity of online payments, skimming of receivables is a diminishing threat in most modern organizations. However, for organizations still employing traditional methods of receiving payment by checks or cash, institution of proper internal controls can prevent or lead to early detection of such schemes. In subsequent audits, identifying unusual patterns on customer accounts could also unravel such schemes.

Use of a Shell Company

High-level employees within an organization with authority over disbursements may create shell companies that they control. These shell companies then bill the organization for fictitious goods and services. The perpetrator usually is in a position to approve charges or has authority over personnel who approve payments on behalf of the organization. As the payment is made to the shell company, the perpetrator has effectively stolen funds from the organization.

Fraudulent shell companies often will use a P.O. box or residential address as a business address. Sometimes the owner of the shell company could be the spouse or other close relative of the perpetrator, and their names or addresses could be used to set up the shell company. Often the billing documents from these shell companies lack the authenticity of legitimate companies. For example, use of a shell company was discovered when a secretary noticed that the street address of a vendor was the home address of her supervisor. In another instance, fraud was revealed when it was observed that invoices from a vendor that were months apart were sequentially numbered. The implication therefore was that the victim organization was the only customer for this vendor. On further investigation, the fictitious vendor was revealed.

Shell companies can at times sell legitimate goods to the company but at an inflated price. The shell company purchases the goods needed by the organization from legitimate vendors and then resells to the organization at an inflated price. The individual(s) who own the shell company pocket the difference. Such schemes are known as pass-through schemes.

Verifying the list of vendors and ascertaining their legitimacy is an effective way of uncovering the use of a shell company. Data analytic techniques could be effectively employed to analyze large amounts of vendor data to identify anomalies and suspicious activities.

The Enron financial scandal increased the public's awareness of the use of shell companies to commit fraud. Even though shell companies were used by Enron for fraudulent purposes, they were not used to embezzle from the company, but rather to falsify their financial statements. Enron's use of shell companies is an example of management fraud where the victim was not the organization but the investors and other third parties.

Ghost Employees

A common fraudulent scheme involving payroll is for Human Resource managers or Payroll managers to create *ghost employees*. The ghost employee, while on the payroll of the company, collects wages periodically but does not actually work for the company. This could be a fictitious person or a family member of the perpetrator. By means of falsifying personnel and payroll records, a ghost employee is added to the payroll and hence collects monthly wages. The potential loss to the victim organization of a ghost employee scheme could be enormous due to the recurring nature of the theft. After the perpetrator has successfully created a ghost employee in the payroll system, the regular process of issuing paychecks ensures a steady stream of funds to the perpetrator. When successfully instituted, unlike the schemes of a shell company or skimming, the perpetrator of a ghost employee scheme does not have to engage in any further maintenance of the fraudulent scheme. As there are no recurring actions on the part of the perpetrator, the data shows no unusual patterns.

The existence of ghost employees is difficult to detect by performing trend analysis or investigating unusual patterns; instead, they can be identified by comparing different databases. The perpetrator could have access to a couple of databases and thus might be able to alter them. However, she will not be able to include the ghost employee in other essential databases to which she has no access. Because the ghost employee doesn't really work at the company, there is no documentation of work performed by this employee, no vacation days taken, no performance evaluation report, and so on. Reconciling employee data across various functions of the organization can help to detect ghost employees. Data mining and statistical techniques are helpful in identifying the handful of employees who are outliers across various organizational functions so the investigation can focus on them.

Inventory Shrinkage

When inventory is sold and the corresponding sale is not recorded (as in skimming discussed earlier) or when inventory is stolen, the perpetrator has to amend the unaccounted decrease in inventory balance. Inventory shrinkage is the reduction in the inventory balance due to theft or waste. Investigating the causes of inventory shrinkage can help unravel fraud schemes. Although some amount of inventory shrinkage is routine and expected in the normal course of business, abnormal shrinkage or a pattern of shrinkage are red flags. Normal inventory shrinkage, a random event, should affect all items of the inventory and not just a particular item. Moreover, there should not be any detectable pattern or trend of inventory shrinkage. Such patterns and trends, if identified through statistical procedures, require further investigation.

Documenting inventory shrinkage can be difficult for many organizations due to their accounting systems for inventory. There are two common methods to account for inventory: the perpetual system and the periodic system. In the perpetual method, every transfer-in and sale of inventory is recorded. On the other hand, in the periodic method, the inventory balance is estimated or computed at periodic intervals. Usually only one of the two inventory systems is used in an organization. To effectively detect inventory shrinkage, a perpetual system has to be implemented to maintain running totals of the inventory that can be verified periodically through physical observation. Discrepancies between the two balances indicate the amount of inventory shrinkage.

Perpetrators have been known to conceal inventory shrinkage by altering either the perpetual inventory records or managing the physical count. A critical internal control procedure, segregation of duties, prevents the perpetrator from altering records to conceal the theft of inventory. For example, an item could be reported as broken or perished prior to its theft by the perpetrator, thus the records are adjusted prior to the actual theft of the inventory item. In the most egregious cases, the inventory items are replaced by empty boxes, giving the illusion of inventory. The fraud case of Crazy Eddie, an electronics superstore in the New York metro area, was an infamous occurrence of inventory padding (see Exhibit 1.1).

Exhibit 1.1   Fraud at Crazy Eddie

Crazy Eddie was an electronic store chain that operated in the New York metro area. It started as a private company in the 1970s and went public in 1984. Although fraud was rampant from its inception, the schemes were modified when the company decided to raise public capital. As a private company it committed insurance fraud and payroll tax fraud. However, in preparation for going public, it primarily exploited timing differences to overstate revenue, inflated inventory to understate cost of goods sold, concealed liabilities and other expenses to understate operating expenses, and resorted to inaccurate disclosures in its financial statements for cover up. As a result of overstating revenue and understating cost of goods sold and other operating expenses, the company reported much higher income than it actually earned. On the flip side, by overstating inventory and understating liabilities, it presented a much more appealing financial position on the balance sheet.

One of the many ways that Crazy Eddie was able to conceal the massive fraud for so long was by developing a tightly knit company culture where most senior employees were either family or friends of the family. Every employee was considered part of the extended family. The initial motivation for fraud, when the company was private, was to pay the employees off the books so that they could evade income and payroll taxes. However, to be able to pay cash off the books, the company needed cash off the books. This was achieved by skimming cash sales and thereby also avoiding paying the sales tax. Thus, by skimming sales to pay employees off the books, Crazy Eddie was able to pocket the 8% sales tax he collected from customers and keep his labor cost lower than his competitions' by paying less to his employees and assisting them to avoid paying income and Social Security taxes.

After the company went public, the objective for committing fraud changed.[8] The family was no longer interested in tax evasion by skimming sales and reporting lower income. Instead they wanted to sell millions of dollars of stock at inflated share prices. To support a high stock price, the company had to find ways to inflate earnings. So the nature of the fraud adapted from understating revenue and income to overstating revenue and income. To inflate reported income, the company not only overstated revenue but understated cost of sales and other operating expenses. The management exploited the auditor's usage of sampling to inflate inventory counts. The employees of the company volunteered to do the auditor's tedious job

> of counting inventory items, which were stacked high and deep in the warehouse. As a result, even though the physical count of inventory was technically being taken, the inventory padding schemes of management were still effective in concealing the massive fraud.

Inventory shrinkage, even when carefully concealed, can be detected by comparing gross margin percentages across various stores. The location that is stealing inventory and reporting it as either sold or spoiled would have an unusually lower gross margin percentage relative to other stores. Furthermore, conducting a trend analysis over multiple periods would lead to early detection of changes in patterns due to inventory theft. Statistical methods, discussed later in the book, can be used to conduct such analysis.

Embezzlement by Management

Three senior officers at Tyco International were convicted of embezzling millions of dollars from the company. The CEO and two other top officials manipulated two corporate loan programs to obtain funds to sponsor their lavish lifestyles and to give themselves unauthorized bonuses. Subsequently, in order to conceal their theft, they would forgive each other's loans and thereby steal the funds from the company.

The formal charges filed against the officers by prosecutors were for stealing $170 million in company loans and other funds and obtaining more than $430 million through the fraudulent sales of securities. The SEC filed a separate but related charge for their failing to disclose the multi-million dollar low interest or interest-free loans they took from the company and in some cases never repaid. It charged the officers with "treating Tyco as their private bank, taking out hundreds of millions of dollars of loans and compensation without ever telling investors."

1.5 Cyber-crime

In a speech delivered at the American Institute of Certified Public Accountants National Conference on Fraud and Litigation Services in 2007, an Associate Deputy Director of the FBI, Mr. Joseph Ford best described the increased threat of financial crimes in the age of globalization and the Internet. He characterized the Internet "as much a conduit for crime as it is for commerce." While opening up new avenues of crime in computer hacking, the Internet facilitates a wide range of traditional criminals that include mobsters, drug traffickers, corporate fraudsters, spies, and terrorists.

The critical challenge that cyber-crime imposes is that unlike traditional crime, the physical presence of the criminal at the scene of the crime is not necessary. In the traditional days of bank robbery, the robber had to be present at the heist and took the risk of being caught and dragged away in cuffs. However, in the age of the Internet, a bank robber could commit a crime of much greater magnitude without ever even having to set foot in the bank. They usually operate from remote locations, under the umbrella of legal protection provided by a different jurisdiction than where the crime is being committed. These criminals exploit the weaknesses in internal controls to commit their crimes. The risk they face is that their plans might be thwarted, but usually the identification, prosecution, or incarceration of such criminals is complicated due to the involvement of multiple jurisdictions. According to the FBI, a significant percentage of cyber-crimes originate in Romania, but the victims are in the United States or in other western nations. These criminals, formerly employed by the now defunct intelligence apparatus of Romania, have targeted money transmission agencies. Additionally, one type of securities fraud, known as a "pump and dump" scheme, has been traced back to criminals in Latvia and Estonia who hack into accounts of online brokerage companies. Although these crimes originate and are committed overseas, the consequences are felt in the United States and adversely impact U.S. investors.

Consequently, the significant threat encountered by banks and other financial institutions is no longer that of an armed robber engaging in a heist, but that of a cyber-criminal sitting in the comfort of his home at a remote location. The weapon of choice for such criminals is not a firearm as used in traditional bank robberies, but a few keystrokes on their computers. Consequently, the analysis of evidence of such crime doesn't require evaluation of fingerprints or ballistics experts. They require a financial analysis of the money trail and hence the expertise of accountants.

1.6 Chapter Summary

Our collective knowledge of fraud schemes and occurrences is limited by those that have been detected and prosecuted with success. Hence, sophisticated fraud schemes currently undetected by auditors, regulators, or law enforcement officials, might be rampant in society, but our current technology and methods are not sufficiently calibrated to detect them. Because we operate under limited knowledge and fraud perpetrators are a step ahead in conniving new schemes, it is imperative for fraud examiners to be fully cognizant of the schemes that have been unraveled and be equipped with tools to prevent and detect such instances in the future. That is, after a fraud scheme has been discovered, the information should be disseminated as widely as possible so that others can take precautionary steps and reduce their vulnerability to such threats. It is toward this objective that this book began with a brief description of commonly used fraudulent schemes by employees, management, and third parties.

Although extant literature and books on forensic accounting have focused on detecting and preventing employee and management fraud in an organization, cyber-crime is a growing threat. Businesses, especially financial institutions, are increasingly vulnerable to cyber-crime. As computing technology makes advances and the use of Internet for legitimate business purposes explodes, it becomes increasingly difficult to monitor Internet traffic and identify suspicious activities. Additionally, though precautionary measures can prevent loss, the perpetrators of cyber-crime are rarely prosecuted with success and therefore are free to continue to devise new schemes to defraud victims. That is, while the perpetrators of a failed robbery would be prosecuted and incarcerated (thereby limiting their ability to perfect their crime), the foiled attempts of cyber-criminals are not prosecuted due to jurisdictional constraints, and these criminals gain valuable knowledge about the cyber-security systems and potentially use it to beat the system. Thus there is now a more critical need to be proactive in developing better and impregnable *cyber-shields* to protect against cyber-crime. The statistical techniques presented in subsequent chapters provide the necessary foundation to develop sophisticated investigative tools to prevent and detect not just employee and management fraud but also to deter cyber-crime.

1.7   Endnotes

1. In its *Financial Crimes Report to the Public*, the FBI summarizes its operations for the two-year period ending on September 2011. The entire report is available at http://www.fbi.gov/stats-services/publications/financial-crimes-report-2010-2011.

2. *Washington Post* in July 2010 estimated the losses to be $21.2 billion, http://voices.washingtonpost.com/economy-watch/.

3. The U.S. Securities and Exchange Commission describes the warning signs of a Ponzi scheme at http://www.sec.gov/answers/ponzi.htm#RedFlags.

4. For a detailed discussion of Lehman's use of Repo 105 transactions, see Caplan, Dennis H., Saurav K. Dutta and David J. Marcinko, "Lehman on the Brink of Bankruptcy: A Case about Aggressive Application of Accounting Standards." *Issues in Accounting Education* 27, no. 2 (2012): 441-459; or Dutta, Saurav K., Dennis Caplan and Raef Lawson, "Lehman's Shell Game: Poor Risk Management." *Strategic Finance*, 2010 (August).

5. Available at http://lehmanreport.jenner.com/. Volume III of the Examiner's Report is particularly relevant to the issues raised in this case.

6. *Principles of Fraud Examination,* Joseph T. Wells, 2010, John Wiley and Sons Publishing.

7. Lapping is one of the most common concealment techniques (*Principles of Fraud Examination* by J. Wells, p. 66).

8. A thorough discussion of the fraudulent scheme at Crazy Eddie is presented in *Contemporary Auditing: Real Issues and Cases* by Michael Knapp, Southwestern Publishing.

2

Legislation, Regulation, and Guidance Impacting Forensic Accounting

You'll have lower prices under deregulation than you will through regulation.
—Kenneth Lay, CEO of Enron

2.1 Introduction

Regulation and legislation are never popular, but sometimes they are essential in preserving societal values. In the wake of the corporate scandals of the early 2000s and the financial crisis of 2007, in which billions of investor dollars were lost, there has been a growing call for stricter enforcement of regulation that addresses white collar financial crimes. Accounting as a profession is an important component and contributor to such financial regulation.

The capital markets in the U.S. function efficiently due to the implicit trust between investors and professional managers. Accounting is an important medium for building that trust. However, on occasions when that trust is violated or compromised through acts of management fraud, investor trust and confidence in the markets have to be restored. In those instances, legislative bodies and regulators must take necessary actions to restore faith in the proper functioning of the capital markets.

The corporate scandals of the late 90s and early 2000s, such as Worldcom, HealthSouth, Enron, Computer Associates, and Tyco among others, prompted actions both by the legislative branch as well as the executive branch of the U.S. government. The legislative response culminated in the Sarbanes-Oxley Act. The executive action led to the creation of The President's Corporate Fraud Task Force with sweeping authority to combat and prosecute corporate fraud. This chapter summarizes those developments and the consequences thereof.

2.2 U.S. Legislative Response to Fraudulent Financial Reporting

The Sarbanes-Oxley Act of 2002 (SOX) was a legislative response to the numerous corporate scandals of the early 2000s. SOX was meant to restore investor confidence in the capital markets, among other things, and improve the quality of financial disclosure and corporate governance by public companies. To encourage accurate financial disclosure, Sections 302 and 404 of SOX mandated management and the external auditor to perform assessments of and to report on the effectiveness of the issuer's internal controls over financial reporting.

Section 302 of SOX requires financial reporting certifications by an issuer's Chief Executive Officer and Chief Financial Officer. Further, it made these officers of the company responsible for maintaining and periodically evaluating the effectiveness of internal controls and required them to list all deficiencies in internal controls as well as list changes in business processes that could have a negative impact on internal controls.

Section 404 requires management and the external auditor to report on the effectiveness of the company's internal controls. Section 404(a) requires management to implement, document, and test internal controls over financial reporting and to issue a report on an annual basis. Section 404(b) requires the issuer's independent auditor to opine on the effectiveness of the issuer's ICFR. The SEC phased in compliance with Section 404(b) for issuers based on filing status.

In 2009, the SEC conducted a study of the requirements under Section 404 of SOX, including the costs and benefits associated with Section 404(b). This study analyzed survey data about these costs and benefits for a wide variety of issuers. One conclusion from this study was that the Section 404(b) compliance costs vary with company size and compliance history. Whereas larger companies tend to incur higher absolute costs, smaller companies report higher "scaled" costs (or costs as a fraction of total assets). Additionally, this study found that the 2007 reforms, which include the SEC's June 2007 Management Guidance and the Public Company Accounting Oversight Board's (PCAOB) Auditing Standard No. 5 (AS 5), have resulted in a reduction in the cost of compliance under Section 404.

In July 2010, the Dodd-Frank Wall Street Reform and Consumer Protection Act (Dodd-Frank) was enacted by Congress. A provision in this Act exempted nonaccelerated filers (generally those issuers with less than $75 million in public market float), who comprise about 60% of all issuers, from the requirements of Section 404(b) of SOX and required the SEC to conduct a study to determine ways that it could reduce the compliance burden of Section 404(b) while maintaining investor protections for issuers with market capitalization between $75 million and $250 million. On April 22, 2011, the SEC published its required study and submitted it to Congress. The study found that investors generally view the auditor's attestation as beneficial and that financial reporting is more reliable when the auditor is involved with the ICFR assessment. The study recommended no further exemptions from the requirements of Section 404(b) for issuers in the studied population or for any other issuers.

In providing further guidance to companies for them to comply with the provisions of SOX, the SEC made the following clarifications:

> The first principle is that management should evaluate whether it has implemented controls that adequately address the risk that a material misstatement of the financial statement would not be prevented or detected in a timely manner. (SEC Release 33–8810, page 4)

In specifically addressing the risk of fraud within an organization, the SEC states

> Management's consideration of the risk of misstatement should include consideration of the vulnerability of the entity to fraudulent activity... Management should recognize that the risk of material misstatement due to fraud ordinarily exists in any organization, regardless of size or type, and it may vary by specific location or segment and by individual financial reporting element. (SEC Release 33–8810, page 14)

The SEC guidance explicitly requires certain types of disclosure when incidents of management fraud are unraveled:

> ...identification of fraud, whether or not material, on the part of the senior management. (SEC Release 33–8810, page 50)

SOX has a significant influence on the design, maintenance, evaluation, and reporting of internal controls for medium to large publicly traded companies. It requires management and the auditor to periodically evaluate the internal controls and identify material weaknesses. Identification and reporting on material weaknesses requires management and the auditor to address those weaknesses and develop plans to overcome them either through better design or implementation of existing controls. The Act also increases reporting responsibilities for both management and auditors.[1]

2.3 The Emphasis on Prosecution of Fraud at the Department of Justice

In its 2012 Annual Report to the American Public, the Department of Justice stated that providing protection from financial and health care fraud is one of its highest priority goals. The other two highest priority goals were providing protection from terrorist activities; and combating gang violence. This elevation in emphasis on protection from fraud was partially triggered by the recent financial crisis and the impact it had on the general American population. The resultant fraud in finance and housing market has generated much pressure on governmental agencies to take appropriate actions and to prosecute individuals who commit financial crimes.

The traditional labeling of white collar financial crimes as being "victimless crimes" has been questioned in the wake of Enron, Worldcom, Madoff, Stanford International Bank, and other financial scandals wherein the life savings of innocent victims have been wiped out as a result of such crimes. Consequently, the Department of Justice has considered it a top priority to prosecute criminals who commit mortgage fraud, securities and commodities fraud, and other types of financial fraud that victimize the American public as a whole.

President Bush created the President's Corporate Fraud Task Force in July 2002 to restore public and investor confidence in America's corporations following a wave of major corporate scandals. Since its formation, the Task Force has fiercely prosecuted corporate fraud and brought to justice those who have committed financial crimes such as accounting fraud, securities fraud, insider trading, market manipulation, money laundering, violations of Foreign Corrupt Practices Act, stock option backdating, and others. In less than a seven-year period it yielded 1,300 fraud convictions including more than 200 chief executive officers and presidents, more than 120 corporate vice presidents, and more than 50 chief financial officers. The significant cases that were tried by this Task Force include

- Enron, in which criminal charges were brought against 36 defendants, including 27 former company executives. The task force seized more than $100 million and recovered more than $450 million for the victims.

- Enterasys, in which eight former officers including the Chairman and the CFO pleaded guilty to artificially inflating revenue to increase the stock price. The company was found to have overstated its revenue by $11 million, resulting in shareholder losses of about $1.3 billion.

- Qwest, in which the former CEO was convicted of insider trading charges stemming from his sale of approximately $100 million in the company's stock. He had been informed of material, nonpublic information regarding the company's financial health.

- AEP Energy Services admitted that its traders manipulated the natural gas markets by knowingly submitting false information to market indices.

- PNC, the seventh largest bank holding company in the U.S., was charged for fraudulently transferring $762 million of troubled loans to off-balance sheet entities.

- Cendant's former Chairman and Vice-Chairman were convicted of an accounting fraud scheme that spanned a decade. The fraud, when revealed, resulted in investor losses of more than $14 billion in one day.

- Mercury Finance, a sub-prime lending company, was charged with inflating revenue and understating its delinquencies and write-offs. The fraud resulted in investor losses of more than $2 billion in a single day.

- Homestore, an Internet company, was convicted of setting up bogus "round-trip deals" to inflate its revenue by millions of dollars.

- Adelphia's CEO and CFO were convicted of fraud charges and embezzling from the company.

- WorldCom's former CEO and other senior officers were convicted of accounting fraud. Details of the fraud were discussed in Chapter 1, "Introduction: The Challenges in Forensic Accounting."

- Refco's former CEO and CFO were charged in a scheme to hide massive losses sustained by the company in the 1990s amounting to more than $2 billion.

- Comverse's former CFO was charged with backdating option grants and granting options to fictitious employees.

- Dynegy's three former executives were convicted of an accounting scheme where they misrepresented proceeds from debt as revenue from operations.

These and many other cases highlight the gravity that federal law enforcement officials in the U.S. have been placing on financial crimes. The following section discusses the role that the FBI plays in investigating and detecting financial fraud that enables the Department of Justice to prosecute the offenders.

2.4 The Role of the FBI in Detecting Corporate Fraud

As of March 2012, about 15% of the FBI's agents qualified as special agent accountants. Accountants have been woven into the fabric of the Agency since its inception in 1908. At this time approximately a third of the original task force of 34 investigators were bank examiners.[2] The critical importance for accounting expertise within the FBI was underscored during the Savings and Loan crisis of the 1980s when accounting technicians were required to help agents sort out complex financial transactions. In the aftermath of the high profile corporate financial scandals of the 2000s and the financial crisis of 2009, the Agency created a specific job title known as the *forensic accountant*.

The responsibilities of a forensic accountant include[3]

- Conduct financial analysis of business and personal records.
- Help identify individuals or groups participating in suspicious or illegal activity by following the money trail.
- Participate in gathering of evidence of a financial nature.
- Accompany case agents in interviewing subjects on topics of a financial nature.
- Identify and help trace funding sources and interrelated transactions.
- Assist prosecuting attorneys in developing strategies.
- Serve as expert witnesses in judicial proceedings.

The qualifications required for a forensic accountant position at the FBI include a bachelor's degree in accounting with preferred one or more relevant certifications, such as

- Certified Public Accountant (CPA)
- Certified in Financial Forensics (CFF)
- Certified Fraud Examiner (CFE)
- Certified Internal Auditor (CIA).

In the previous chapter a detailed discussion was presented on the various initiatives that the FBI is currently undertaking to combat fraud in every sphere, from corporate to health care.

2.5 Professional Guidance in SAS 99

A white paper published by the American Institute of Certified Public Accountants (AICPA) as an addendum to the Statement of Auditing Standards No. 99, *Consideration of Fraud in a Financial Statement Audit*, provides suggested anti-fraud programs and controls. This documentation provides a linkage between internal control deficiencies that perpetuate fraudulent actions. It also provides guidance to companies in setting up appropriate oversight processes. Finally, the document suggests some benchmarks to evaluate the adequacy of fraud programs and controls. The document also includes two templates on code of conduct and ethics that could be adopted by organizations seeking to cultivate a culture of ethical behavior in their organizations.

The AICPA defines fraud very broadly to encompass minor employee theft and unproductive behavior to misappropriation of assets and fraudulent financial reporting. The Statement of Auditing Standards document is based on the presumption that the risk of fraud can be reduced, if not eliminated, through a combination of prevention, deterrence, and detection mechanisms. While acknowledging that fraud may be difficult to detect due to the efforts of the perpetrator to conceal the theft through falsification of documents or collusion amongst employees, it emphasizes the importance of prevention mechanisms. The document reasons that prevention and deterrence are more efficient than the consequent detection and investigation measures. It encourages organizations to instill organizational core values and principles throughout the organizational decision making process so that it guides actions of all employees.

The document asserts the organization's responsibility to create a culture of honesty and good ethics and to communicate benchmarks of acceptable behavior to each employee. Such a culture should be rooted on a strong foundation of "core values." Promoting an ethical culture throughout the organization leads to other indirect benefits of creating a good work place environment and enhances the organization's ability to attract and retain good employees.

In creating a culture of honesty and high ethics, consideration of the following are suggested:

- Setting the tone at the top. This recommendation is grounded on research in moral development that suggests that "action speaks louder than words" when it comes to urging ethical behavior on the part of the subordinates.

- Creating a positive workplace environment. Employees are less likely to cheat and defraud organizations when they have a positive feeling toward it. Poor employee morale is linked to higher incidents of undetected and unreported cases of employee fraud.

- Hiring and promoting appropriate employees. For example, conducting background checks prior to hiring employees reduces the risk of hiring individuals with criminal records.

- Training and confirmation. Newly hired employees should undertake mandatory training on the organization's core values. The organization's expectations of ethical behavior should be unambiguously communicated to the employees.

The document also identifies three steps that organizations should undertake to evaluate antifraud processes and controls. These steps are

1. Identification and measurement of fraud risks. The document suggests considering fraud risk as part of an enterprise-wide risk management program, and the organization's vulnerability to fraudulent activities should be evaluated.

2. Steps to mitigate fraud risks. This requires management to consider changes to activities and processes that might reduce or eliminate the risk of fraud.

3. Implementation and monitoring of appropriate internal controls.

The AICPA guidance formulated in conjunction with other member organizations provides a roadmap to implementing and maintaining ethical culture and adequate internal controls to prevent and deter fraudulent activities throughout the organization.

2.6 Chapter Summary

This chapter provided an essential but brief overview of the legislative and enforcement processes affecting corporate reporting. The material presented conveys the magnitude and impact of such crime. The financial losses resulting from corporate fraud approach billions of dollars in lost investor wealth. Additionally, the numbers of victims affected by such financial crimes are large and belong to every strata of the society. The employees, pensioners, retirees, and others who invest in these companies through pension funds require some level of protection that their livelihoods and lifetime savings will not be wiped out by reckless actions of a few.

The government and the regulators have responded to the public sentiment by enacting legislation that improves corporate governance at publicly traded companies. Although the companies could voluntarily implement such measures to further their economic value, some had not done so on their own. The requirement of designing and maintaining an adequate system of internal controls, while beneficial to all businesses, was not fully undertaken by some at great peril to ordinary investors and creditors. Additionally, the requirement of disclosing threats and failures of internal controls prompt management to take these violations seriously and take steps to rectify observed weaknesses.

2.7 Endnotes

1. *Complying with Sarbanes-Oxley Section 404* by Lynford Graham provides a detailed guide on the implementation issues related to SOX.

2. The information available at FBI's story published in March 2012 titled, "FBI Forensic Accountants Follow the Money," http://www.fbi.gov/news/stories/2012/march/forensic-accountants_030912.

3. The role and responsibility of forensic accountants employed by the FBI is summarized on the Agency's website and available at http://www.fbi.gov/news/stories/2012/march/forensic-accountants_030912.

3

Preventive Measures: Corporate Governance and Internal Controls

The problems we have cannot be solved by thinking the way we thought when we created them.
—Albert Einstein

3.1 Introduction

There is widespread public recognition, supported by substantive academic research, that corporate governance is essential to protecting shareholder interests and is critical to the sustainability of a firm. Many large business failures have been attributed to lax corporate governance within firms that were ineffective in preventing the downfall or informing various stakeholders on a timely basis. Corporate governance is often defined as the system and processes by which organizations are directed and controlled. It deals with the alignment of the interests of corporate managers with those of other stakeholders, primarily owners or shareholders.

Economic models of manager behavior, commonly known as the *agency models*, provide theoretical underpinnings for corporate governance features. The basic premise of the agency model is a separation between the ownership and the management function.[1] Thus, the interests of the managers and those of the owners might not be perfectly aligned. This divergence in the interests can cause managers to take actions that are costly to the owners. Such activity cannot be precluded based solely on contracts, as owners are unable to directly observe the manager's actions or quality. Further, extensive monitoring as well as comprehensive contracting are prohibitively costly. This results in two well-studied problems: *moral hazard* and *adverse selection*. In this context, the features of corporate governance can be interpreted as a characteristic of the contract that governs relations between owners and managers. Challenges to governance arise due to the same unobservable features of managerial behavior or ability.

An alternative to direct oversight by the shareholders is governance through a Board of Directors (BOD or Board) who are elected by the shareholders. Although the Directors' interests might not fully overlap with those of the shareholders, they are better aligned. However, because management might have some degree of control over the selection of the Directors and can also serve on the BOD, it could compromise the underlying intent. Consequently, it is widely recognized that the greater the independence and autonomy of the BOD, presumably, the higher the level of corporate governance.

The agency literature has been mostly developed and studied in the context of developed economies and western culture. Some of the results and insights are applicable to other economies and cultures, but some are not. Recognition and understanding of these similarities and differences from a global standpoint is important when applying these concepts to other settings and cultures. In Sections 3.2 and 3.3 the principles of the agency theory are presented both in the context of developed western economies and developing eastern economies.

Four key features of this chapter are summarized in Figure 3.1. We begin by discussing corporate governance challenges in deverloped and emerging economies in sections 3.2 and 3.3, respectively.

Figure 3.1 Four Key Features

Corporate governance and control measures sometimes act as inhibitors to corporate innovation and growth. The organizations seek to obtain the right balance among the diverse objectives. The pervasive topic of organizational controls is summarized in Section 3.4. Sections 3.5 through 3.7 present the important role of internal controls in all organizations. Internal controls, including the definition and common mechanisms, are discussed in Section 3.5. The COSO framework, which was developed in 1992 and has gained wide acceptance, is summarized in Section 3.6. The cost-benefits of internal controls are discussed in Section 3.7. The inherent limitations of internal controls are also discussed in that section. The specific role of internal controls in preventing fraud is discussed in Section 3.8. Section 3.9 briefly summarizes U.S. legislation on companies' having a properly designed and functional internal control system as well as related reporting. Section 3.10 concludes with a summary of important learning objectives presented in this chapter.

3.2 Corporate Governance Issues in Developed Economies

In modern corporations, managers own a relatively small percentage of the firms they manage. In a 2000 survey of U.S. corporations, of a representative sample of large public firms, 90% of the CEOs owned less than 5% of their companies.

This relatively negligible ownership by managers can create problems with goal congruence, which is known in academic literature as *moral hazard*. Moral hazard is defined as a manager's aversion to work and incongruence of goals when a manager's and an owner's interests are not perfectly aligned. Corporate governance mechanisms are developed to prevent such opportunistic behavior through all levels of the organization. Two key mechanisms that have been developed to monitor the actions of the CEO and other top managers of the firm are an independent Board of Directors and a required external audit for all publicly held companies. The BOD is most commonly described as a formal link between the shareholders of a firm and the managers entrusted with the operations of the firm.

An independent BOD is believed to be more capable of monitoring managers because there is little conflict of interest. Empirically, it has been demonstrated that the higher the independence of the BOD, the lower the incidence of accounting fraud and earnings management.[2] Most worldwide regulations on corporate governance are therefore based on the premise that independent directors who have little or no other affiliation with a firm are more effective in monitoring management. In the United States, the Sarbanes-Oxley Act (SOX) of 2002 and the more recent Dodd-Frank Act (2010) legislate based on this premise. In the United Kingdom, the Cadbury Report, issued in 1992, has similar recommendations: at least three independent non-executive directors; separation of the role of Chairman and CEO; and a requirement to have an audit committee and a remuneration (compensation) committee.

Typically, the Board of Directors is responsible for the hiring, compensating, and firing of the CEO of a firm. They also typically oversee the firm's overall business strategy, which includes monitoring its corporate control.

Although the BOD oversees management actions on all facets of the organization, the Audit Committee of the BOD oversees the financial reporting process and accounting issues of the firm. The Audit Committee also oversees the firm's external audit process and its internal controls over financial reporting. In 1999, the Blue Ribbon Panel sponsored by the NYSE and NASDAQ made recommendations about the independence of audit committees. In response, the NYSE started requiring each firm to have an audit committee comprised solely of independent directors, while NASDAQ required that independent directors form a majority of the Audit Committee. SOX also requires formal disclosure by a firm if its audit committee is non-existent, inactive, or is not comprised of independent directors.

In addition to independence, the Audit Committee members are expected to possess financial expertise. Although independence is an important characteristic, without competency it is of limited value. Hence, by the end of 2003, all major U.S. stock markets (NYSE, NASDAQ, and AMEX) started requiring that all members of the Audit Committee be financially literate and at least one member have financial expertise. This requirement is based on the premise that individuals with no literacy or experience in accounting or finance are less likely to be able to detect problems in financial reporting.

A CEO's influence on the BOD can sometimes adversely affect its effectiveness. The greater a CEO's influence on the BOD, the less likely the Board is to suspect irregularities or further independently investigate those suspicions. Concerns about a CEO's influence on the Board have led the NYSE to propose that each BOD have a nominating or corporate governance committee that is comprised solely of independent directors. The NYSE views BOD nominations to be among the more important functions of a Board. Hence, an independent nominating committee can further enhance the quality of appointed BOD members.

In essence, key corporate governance mechanisms in the U.S./U.K. are geared toward reducing the influence and power of the CEO and the control that top management exerts on the Board of Directors. Through a distribution of power and corporate decision making, it is presumed that the principal-agent problem shown in Figure 3.2, Panel A, can be mitigated and the decision making would be better aligned with the interest of the shareholders. Specifically, the influence of the "below the surface" relationship between management and the Board is lessened.

3.3 Emerging Economies and Their Unique Corporate Governance Issues

Emerging economies are usually defined as low-income, rapid growth countries that have recently undergone an economic liberalization policy.[3] Typically these economies do not have an effective and predictable legal system, which fosters a weak corporate governance environment. Some of the emerging economies have attempted to adopt legal frameworks similar to the Anglo-American system, either as a result of internally driven reforms (for example, in China or Russia) or as a response to international demands (South Korea, Thailand). However, formal institutions that deal with laws and regulations related to securities trading and information disclosures are either absent, inefficient, or ineffective.[4] Additionally, alternative informal arrangements are prevalent in these economies. These arrangements include relational ties, business groups, family connections, and government contacts that play a greater role in shaping corporate governance.[5]

In emerging economies, the agency conflict is different than it is in the developed economies of the U.S. and the U.K., where conflict is described as a *principal-agent problem*, that is the dissonance of goals between the managers and the owners. In emerging economies, the conflict is between the two types of owners, the controlling shareholders and the minority shareholders. In academic research this is termed as the *principal-principal problem*. The differences and the implications are graphically illustrated in Figure 3.2. In Panel A of the figure, the principal-agent conflict is depicted by the solid line, and the corporate governance measures are depicted by the dashed lines. The dotted lines denote the "under the surface" relationship that exists among various parties that undermines the corporate governance measures. For instance, the solid line connecting the shareholders and the agents depict the principal-agent problem; the dashed lines connecting the shareholders with the BOD and the BOD with external auditors depict the corporate governance measures. However, the dotted line connecting the auditors to management depicts the potential conflict of interest on the part of the auditors as they provide non-audit services to management. Similarly, the dotted line between management and the BOD suggests the inappropriate control that top management might have on the BOD by influencing selection or removal of individual Directors to and from the Board.

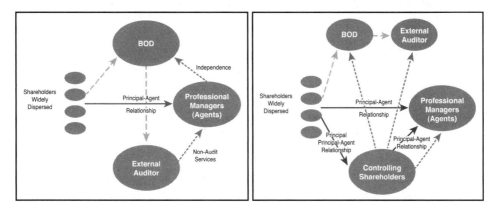

Figure 3.2 Variations in agency problems across cultures and economies

In developed economies, there is usually only one group of shareholders that need not be further fragmented; in emerging economies, the most significant player in the corporate world is the controlling shareholder. This conflict between controlling shareholders and minority shareholders is depicted in Panel B of Figure 3.2. This conflict is depicted by the solid line connecting the two parties and is labeled a *principal-principal* relationship. The presence of controlling shareholders creates new compromises to the traditional corporate governance measures depicted by additional dotted lines connecting the controlling shareholders to the BOD and management. Ironically, when corporate governance measures developed to counter principal-agent conflicts in Western economies are implemented in developing economies, they result in further strengthening the controlling shareholders at the expense of the minority and dispersed shareholders. An interesting case of sibling rivalry that led to the break-up of Reliance Industries, a public company, is discussed in Exhibit 3.1. This conflict often results in expropriation of value from the minority shareholders to the controlling shareholders. Some of these firms from emerging economies register in the U.S. exchanges through American Depository Receipts (ADRs), thus they fall under the purview of the U.S. Securities and Exchange Commission, and the minority shareholders in this case are the U.S. investors. Hence, it is important for U.S. auditors and accountants to be cognizant of this difference in business culture and the associated risks in financial reporting to help safeguard the interests of minority shareholders.

Exhibit 3.1: Sibling Rivalry at Reliance Industries

Reliance Industries, one of the largest conglomerates in India, has recently been through a family feud due to sibling rivalry after the death of the patriarch of the family, Dhirubhai Ambani. The founder had a modest beginning but built an empire in textiles through his savviness and tenacity.[6] When Dhirubhai Ambani died in 2002, leaving no will, Reliance was one of India's largest companies. But it was not big enough for his two sons, who started squabbling. The company was divided into two parts by their mother. However, the family feud still continued five years after the "de-merger" and separate existence.

Since the 2005 de-merger, each segment of Reliance Industries has blossomed. As of 2011, RIL, the conglomerate active in energy, refining, and petrochemicals, run by Mukesh Ambani, was worth $55 billion, and the Ambani family and friends own 45% of it. The two brothers, Mukesh Ambani (MBA, Stanford) with net worth of $29 billion and Anil Ambani (MBA, Wharton) with net worth of $13.7 billion, have been in disagreement over the price of natural gas that Mukesh's company drills and

> Anil's company purchases to run his power plants. In 2010 the brothers took their fight public, when Anil purchased front page advertisements in a leading newspaper accusing the Government of India of favoring his brother's conglomerate. Their disagreement was over a 2005 agreement brokered by their mother that allowed Anil's company to buy gas from the "Krishna-Godavari basin," owned by Mukesh, for 17 years at a price lower than that set by the government. Mukesh had the "deal" annulled in a legal fight, and the case reached the Supreme Court of India. The Court ruled in favor of Mukesh, arguing that the natural resources belong to the nation and cannot be subject to a family pact and sold at cut-rate prices. However, this saga brings to attention the important role the "family" and "matriarch" plays even in the largest businesses in developing economies, in addition to "seemingly legal" arrangements of selling products at below market prices to affiliated companies.
>
> According to *The Economist*[7] each would like to see the other fail. However, despite their corporate fights and legal actions, the two brothers Anil and Mukesh Ambani live in the same high-rise family mansion in Mumbai with their mother, albeit on different floors. The "warring" brothers briefly suspended their hostilities and hosted a party for their mother's birthday in February of 2010 in the midst of the legal scrambling.

Firms in emerging economies that have a controlling family or business interest face many risks, as discussed in Exhibit 3.1. These firms tend to exhibit nepotism in hiring and tend to hire less-qualified family members, friends, and cronies for key positions.[8] This is because to the majority shareholder, loyalty and trust could be more critical than competency and ability. These firms might purchase products and services from other affiliated companies at above market prices or sell goods and services at below market price to organizations owned by or associated with controlling shareholders. These seemingly related party transactions might not be adequately disclosed as such disclosures might not be legally mandated or such transactions are the social norm. These firms may also engage in strategies or initiatives that work to further personal, family, or political agendas of the controlling parties (families) at the expense of firm performance.[9] Sibling rivalry, generational envy, nonmerit-based compensation, and irrational strategic decisions are additional risks that can affect even the largest of firms in emerging economies.[10] Additionally, the relationships between controlling principals (family owners) and agents (family-member managers) are based on emotions, sentiments, and informal linkages that can make it difficult to objectively assess a manager's performance. This is of particular importance in countries where culture places a high value on family relationships, as is implicit in the following quote from the Chairperson of the Securities and Exchange Commission of the Philippines:

There is a lack of transparency in board action and management (of family corporations) since families do not feel the need for public disclosure. As a result, minority shareholders are often kept in the dark as to the actual status of the corporations of which they are part-owners because the large shareholders dominate decision-making activities involving the company.[11]

As was shown in Figure 3.2, the ellipse denoting "controlling shareholders" have a dotted line to management, the BOD, and the external auditors, thereby controlling all of the traditional corporate governance checks at the expense of the minority shareholders. For instance, controlling shareholders might have veto power over who gets selected to the BOD, thereby effectively nullifying the BOD's ability to oversee controlling shareholders and their related party transactions with the firm. Similarly, the dotted line between the controlling shareholders and the professional managers (agents) signify the direct control that the controlling shareholders exercise over professional managers in this setting, thus blurring the distinction between the firm's interests and the personal interests of the controlling shareholders. The case of the acquisition of Vanda Computers in Hong Kong (Exhibit 3.2) is an illustration of how one majority shareholder can exploit minority shareholders.

Exhibit 3.2: Minority Shareholders at Vanda Computers

In 2004, Hutchison Whampoa Ltd. began unloading shares in the affiliate Vanda Systems and Communications Holdings in Hong Kong. Hutchinson Ltd.'s Chairman Li Ka-Shing was the richest man in Asia and is nicknamed "Superman" based on his reputation as a master trader. In the sale, Hutchinson earned a profit of HK $1.3 billion (US $167 million). Hutchinson Whampoa acquired a 30% stake in Vanda a week prior, and in the wake of the transaction Vanda's share price rose 61% in one day. Vanda's shares subsequently fell by 31% after Hutchinson sold Vanda shares just one week after purchasing them. Such a large profit over such a small period might be considered a borderline fraudulent "pump and dump" scheme that would lead to further investigations in some developed markets like the U.S.

There are many structural reasons why controlling ownerships are more prevalent in emerging economies. Primarily, relinquishing controlling ownership requires the sharing and divulging of sensitive information. Without the legal protection and requisite safeguards to intellectual property, founder/dominant ownership is more hesitant to do so. Additionally, the core competencies for firms in emerging economies are still in their developmental stages, and divulging such strategic information could be detrimental to their growth. Further, sharing of information with third parties requires that there be some degree of trust between the parties in that the receiving party will use the information only for the agreed upon intent and not to exploit the divulging party. However, institutions that foster trust between unrelated parties and thereby protect both parties are in their formative stages in emerging economies and might lack adequate resources to be effective.

In summary, dominant ownership is the norm in even the largest corporations in emerging economies. In such economies, not only is ownership more concentrated, but controlling shareholders are likely to be holding more than 50% of firm equity. The controlling shareholder is often associated with a family and/or a business group. The controlling shareholders thus have a motive and means to exploit their positions at the expense of the firm. Such motivation and ability creates additional challenges for external auditors and forensic accountants.

3.4 Organizational Controls

As organizations increasingly grow larger and become more complex and decentralized, it is becoming more challenging for top management to control and synchronize various business functions across multiple locations. Further, as organizations expand their businesses globally, they are increasingly operating in multiple countries with diverse cultures. With such expansion in worldwide operations and hiring of employees from diverse cultures, it becomes even more important to have a pervasive organizational culture that transcends the regional and cultural differences. The organizational credo and the tone at the top are no longer mere slogans but critical to harmonizing diverse interests into unified organizational objectives.

Additionally, with increased global competition, organizations face a daunting task of continuous improvement and innovation. Improvements and innovation mandate change. However, control systems and standard operating procedures inhibit change. This causes organizations to address an urgent need to better balance the flexibility and nimbleness needed for innovation with the increased vulnerabilities to their reputation, and perhaps their existence. This dilemma of balancing these seemingly disparate objectives is a vexing problem for present-day organizations. Robert Simons in his 1995 book *Levers of Control* proposed four interacting and compensating measures to balance these disparate objectives, as shown in Figure 3.3. These levers are the belief system; a boundary system; a diagnostic control system; and an interactive control system.

The *belief system* communicates the company "credo" or core values to the employees. This is usually achieved by adopting a mission statement. A mission statement helps to communicate throughout the organization the primary objectives of management and the direction they want to take. Every organization is created with a goal or objective. These goals or objectives can change in order to adapt to a changing societal norm. Traditionally (in Milton Friedman's words), the objective of a public company is to maximize shareholder wealth; however, recently organizational goals are usually broader than that and encompass social responsibilities in addition to the profit maximizing objective.

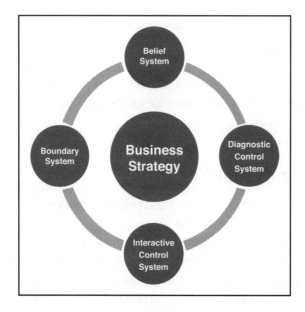

Figure 3.3 The Four Levers of Corporate Control

(Adapted from *Levers of Control* by Robert Simons)

A "boundary system" as the name implies sets limits on employee actions. The boundary system is more flexible than traditional rules or standard operating procedures and hence empowers employees to be creative and entrepreneurial within certain broad limits. Although a belief system is the positive lever that encourages employees to be entrepreneurial and innovative, the boundary system imposes restrictions on such initiatives. Thus, it assumes a negative role in an organization. A boundary system might be easier to implement across international boundaries, rather than enforcing standard operating procedures across diverse cultures. For multi-national companies, aligning standard operating procedures across international boundaries can be difficult due to cross-border differences in laws and regulations. These cross-country differences not only exist in financial and accounting regulations, but more importantly in environmental, social, and safety norms.

A diagnostic control system is akin to standard variance analysis in traditional management accounting. Actual results are compared relative to planned performance measures or budgets. Keeping track of "key" performance measures on a timely basis is critical to managing a business and quickly adjusting to deviations or market trends. The factors attributable to failure or excessive risk become the source of identifying critical performance variables.

An interactive control system enables top-level management to keep track of critical activities that demand frequent and regular attention. These systems are used to focus organizational attention on strategic uncertainties and provide a means to adapt strategies in response to changes in the business environment. There are many other similar management tools, such as "dashboards" that achieve similar objectives.

Although empowerment and innovation are sanctioned by the beliefs and interactive control systems, the constraints are provided by the boundary and diagnostic control systems. In some sense, beliefs and interactive control systems can be viewed as forward momentum, and in contrast, the boundary and diagnostic systems are forces of friction that impede growth by setting limits to actions that may be undertaken. Effective organizations learn how to harness all four levers and propel sustainable growth.

3.5 A System of Internal Controls

Internal controls are processes instituted within an organization to provide reasonable assurance on the following:

- **Safeguarding of organizational assets:** Assets belonging to an organization have to be deployed to further the interest of the organization and not that of a separate entity including its managers or employees. With the advent of the information age, business information has become a critical asset, and safeguarding company secrets and intellectual property is of critical importance.

- **Accurate recording of "only and all" legitimate transactions:** Accurate recording of each and every business transaction is fundamental to the integrity of financial reporting. For instance, if all sales transactions are not recorded, that would understate revenue and may lead to misappropriation of cash generated from the sale. Similarly, a failure to record expenses would lead to an overstatement of income. On the other hand, recording fictitious sales increases income and assets and is a commonly perpetrated fraudulent reporting scheme.

- **Providing reasonable assurance that the accounting rules for the jurisdiction (IFRS, US GAAP or other) are being adhered to:** Accounting rules dictate when and how a financial transaction should be recorded. For instance, recognition of revenue requires some preconditions be satisfied, hence the internal controls have to ensure that the associated journal entry related to revenue recognition cannot be made prior to meeting those preconditions. Absence of such a control, while not necessarily implying misstatement, does increase the likelihood of it.

- **Complying with applicable laws and regulations:** Violation of applicable laws and regulations can have severe consequences for the company. It not only bears a financial cost in terms of fines and penalties, but an adverse reputational effect as well. More importantly, violations of regulations and laws are unethical, and the corporation should institute internal controls to prevent actions that violate applicable laws and/or regulations, such as safety law, child labor law, the foreign corrupt practices act, and other laws.

- **Promoting operational efficiency:** Internal controls prevent waste of resources thus promoting efficiency. For instance, internal control procedures such as supervision increase productivity of workers as the workforce is better utilized with less duplication of efforts.

- **Adherence to management policies:** Internal controls are a means of setting boundaries on employee actions thereby ensuring adherence to management policies.

Traditionally, internal controls are conceptually classified into three categories based on their purpose: preventive controls, detective controls, and corrective controls. *Preventive controls* are designed to prevent errors from occurring. *Detective controls* are designed to detect errors that have already occurred. *Corrective controls* rectify the errors that have occurred. However, these are not mutually exclusive. A particular control could perform two or even all three of these functions. An electrical fuse is an everyday example of a preventive control; when the electricity surges, the fuse melts, preventing additional damage to expensive electrical appliances. A quality control department is an example of detective control; it takes place after the production, hence cannot prevent defects from happening but will detect them on a timely basis. A rework facility is an example of a corrective control.

Subsequent to the introduction of computer technology to the accounting process, the traditional control framework was broadened in the 1990s to include additional information processing controls. These are broadly categorized in two categories: general controls and application controls. *General controls* apply to an entire information system and include features such as emergency lock down procedures, backing up of data, disaster recovery procedures, and so on. *Application controls* are associated with a specific program or application, such as a purchasing system or an employee payroll system. For example a field for social security number will allow data that is exactly nine digits.

Internal controls are designed specifically to meet the needs and objectives of each organization. There is no single internal control design that would be appropriate for all organizations; hence, a comprehensive list or even a requisite list of internal controls is not possible. Some of the more common internal control mechanisms are

- **Segregation of duties:** Segregation, or separation, of duties implies that different employees should perform different tasks that are incompatible from an accounting control perspective. Primarily the three functions that are segregated are authorization, custody, and recordkeeping. That is, the individual responsible for custody of the asset should not be entrusted with recordkeeping, and vice versa. That way, all are prevented from stealing assets and altering the records to hide the theft.

- **Numbered documents:** Important accounting documents such as checks, purchase orders, sales invoices, and others are always prenumbered to ensure that all transactions are recorded and gaps in the sequence can be easily investigated. An out-of-sequence document raises a warning sign and requires further investigation.

- **System/process documentation:** Knowledge of the procedure(s) is essential to assessing the effectiveness of any system including internal controls. Documentation for internal controls is usually prepared as flowcharts or data flow diagrams. This enables employees to examine, critique, and identify weaknesses in the system and determine whether they are functioning as intended.

- **Reconciliation:** Reconciling accounts at a regular frequency is an important aspect of control. It enables timely detection of errors and instituting corrective measures if needed. It also provides assurance, on a regular basis, that the system is functioning as intended and prevents large losses from accumulating.

- **Batch control totals:** When processing a group or batch of a large number of documents, it is customary to calculate various batch totals. Although these totals have no meaning or application, it serves as a check that all records are processed at each interval as the batch moves through the system.

- **Edit checks:** The system "echoes" or repeats the data for verification prior to final processing. Most online transactions, including purchasing airline tickets or hotel rooms, have a confirmation screen prior to final processing of payment.

- **Document matching:** Document matching helps ascertain that vendor invoices are paid only when merchandise has been received and was ordered by authorized personnel. Documents from different divisions would come independently to the disbursement, and cash is disbursed only after all requisite documents are in place. For example, the vendor sends the invoice, the purchasing department sends the purchase order noting the quantity and the price, and the receiving department sends the receiving document noting the quantity. The three independently generated documents can then be matched for conformity.

- **Lockbox system:** This is a commonly used internal control mechanism to prevent theft of cash or checks received by the organization. Typically, businesses require customers to send their payments to a P.O. Box or Lockbox.

- **Physical security:** Physical security over tangible assets through traditional locks is one of the oldest internal control procedures. Simply stated, lock the door and secure the valuables.

- **Back-up procedures:** In the information age, back-up of computer data as well as power supply is of critical importance. Most businesses implement hourly or daily back-ups of accounting data. This prevents the possibility of recreating files in case of a hardware failure. Similarly, back-up power is essential in keeping the information system functioning in case of power outages.

- **Data encryption:** In the age of Internet commerce and wireless communication, data encryption is critically important. Computer hackers and other criminals steal passwords or access to critical data, compromising data integrity and privacy.

- **Firewalls:** Along with data encryption, a firewall is an important element of information system security. A firewall prevents unauthorized intrusion into a system through the Internet. It also cautions the user on the identity of the intruder when an intrusion is detected.

The internal control assessment is a critical part of internal and external audit and was well accepted in the U.S. and elsewhere long before SOX. However, prior to SOX the external auditor had the flexibility of relying less on internal controls and performing additional substantive procedures to obtain assurance on the fairness of the financial statements. That is, the test of internal controls by external auditors, although highly recommended, was not required in all instances. Consequently, there was a limited depth of the standard assessment of the control environment. The SOX act increased the depth of such assessment and broadened the responsibility of auditors to detect and report on material weaknesses in internal controls. The following section discusses these additional requirements imposed on U.S. audits by the Sarbanes-Oxley Act and the consequent standardization of the COSO framework on internal controls. The example of internal control failures at Groupon was publicly disclosed by its auditor Ernst and Young. The details of material weaknesses in Groupon's internal controls are discussed in Exhibit 3.3.

Exhibit 3.3: Groupon's Material Weakness in Internal Controls

Groupon, a Chicago based Internet-coupon company, was launched in November 2008, and in about three years (November 4, 2011) went public with a market capitalization of $13 billion. It had increased its annual revenue from $14.5 million in 2009 to more than $1.6 billion in 2011. In less than six months from its public offering Groupon reported material weakness in its internal control and announced that it had "begun taking steps and plans to take additional measures to remediate the underlying causes of the material weakness, primarily through the continued development and implementation of formal policies, improved processes and documented procedures, as well as the continued hiring of additional finance personnel."

Even before its initial public offering in 2011, Groupon had problems with how it presented its financial results. In one filing, it used a metric called "adjusted consolidated segment operating income" that gave a highly favorable view of its income by not expensing the amount it spent on advertising and acquiring new subscribers. Advertising expenses, which lead to new customers or subscribers, are expected to be expensed as a period expense, thereby reducing reported net income in that period or quarter. However, Groupon reasoned that because a subscriber would stay with the company for a longer time period and make purchases, the subscription acquisition cost could be capitalized and expensed over the expected number of years the subscribers would stay on Groupon. This led to deferral of expenses leading to a higher reported net income and a resultant increase in assets on the balance sheet.

Groupon's revenue recognition method is also controversial even though its business model is simple. It sells a voucher to a subscriber and collects the cash from the subscriber. When the subscriber uses the voucher with the merchant, Groupon gives most of the cash to the merchant, keeping a small amount as commission. Groupon categorized the entire proceeds from the subscriber as its revenue. Some contend that only the commission, or what it collected less reimbursed to the merchants, is Groupon's revenue. In other words, the amount of money Groupon retains after paying the merchant would be called revenue; instead Groupon called that "gross profit." Groupon justified its accounting choice on the basis that it offers full money back guarantee to consumers and not just for the portion it will eventually keep. The company claims that "Groupon" is the product, and sales of that product constitute revenue.

As part of the fallout from the use of controversial accounting metrics and material weakness in its internal controls, Groupon fired its CEO Andrew Mason in March 2013. In his final letter to Groupon employees as the CEO of the company, Andrew Mason wrote:

> "After four and a half intense and wonderful years as CEO of Groupon, I've decided that I'd like to spend more time with my family. Just kidding—I was fired today. If you're wondering why…you haven't been paying attention. From controversial metrics in our S1 to our material weakness to two quarters of missing our own expectations and a stock price that's hovering around one quarter of our listing price, the events of the last year and a half speak for themselves. As CEO, I am accountable."[12]

3.6 The COSO Framework on Internal Controls

In 1992, the Treadway Commission's Committee of Sponsoring Organization (COSO) released a landmark study titled Internal Controls-Integrated Framework. In 2002, the Sarbanes-Oxley Act legislated that internal controls over financial reporting requires an evaluation based on a "suitable, recognized control framework established by a body or group that has followed due process procedures, including the broad distribution of the framework for public comment." This required identification of a "benchmark" against which to compare the effectiveness of any given system. The Securities and Exchange Commission determined that the only body that clearly meets the test is COSO, whose "Integrated Framework" thus became the de-facto standard on internal controls. Although the COSO framework was slowly gaining popularity and was being implemented in practice, enactment of SOX accelerated the process and led to a broad acceptance of the framework in the U.S. as well as internationally. It is now widely recognized as a leading framework for designing, implementing and evaluating the effectiveness of internal controls. This updated framework was released in 2013 to better reflect the changes in business technology and increased stakeholder awareness over the past 20 years. COSO defines internal control as follows:

> Internal control is a process, affected by an entity's board of directors, management, and other personnel, designed to provide reasonable assurance regarding the achievement of objectives in the following categories:

- **Operations Objective:** Affecting the effectiveness and efficiency of operations
- **Reporting Objective:** Enhancing the reliability of reporting
- **Compliance Objective:** Encouraging adherence with applicable laws and regulations

Important aspects in this definition are that internal controls are a means to an end and not the end itself. For example, proper financial reporting is an end (or objective), however the internal control over financial reporting ensures achievement of that objective but itself is not the objective. It is a process that is affected by people, hence the process is only as good as the competency of the individuals operating it. This emphasizes the need to hire qualified and competent personnel to manage these functions. In the absence of competent personnel, a properly designed internal control system is of little value. Finally, effectiveness of internal controls is a probabilistic concept, in that a properly designed internal control does not provide absolute assurance. There is an explicit acknowledgement of risk and uncertainty inherent to the process. Good internal controls positively affect the probability that the desired outcome would be attained. However, the outcome is subject to human error or the possibility that the system might be compromised through collusion of various key employees. Similarly, a poorly designed internal control system increases the probability that errors or material misstatements could be present but doesn't imply for sure that the financial statements are erroneous.

As illustrated in Figure 3.4, COSO defines five major components of internal controls:

- **Control Environment:** The foundation for all other components of internal controls. This includes the "tone at the top," which underscores the importance of internal controls and ethical conduct throughout the organization. It provides the processes, structure, and discipline for internal controls.

- **Risk Assessment:** Involves the continuous and iterative process of identifying and analyzing risks pertinent to the objectives of the organization. The risks to an organization come from within the organization, outside of the organization, and through macro-economic factors or global trends. The assessment also includes the basis for determining how those risks could be managed and having a plan of action should one of those risks materialize.

- **Control Activities:** Activities established at management's directive to mitigate risks that could affect achieving the organization's objectives. These activities are performed throughout the organization and at various stages.

- **Information and Communication:** Generation of critical information and communication of it to internal as well as external parties is a critical function.

Communication of objectives to higher and lower-level employees ensures that all personnel understand their individual internal control responsibilities and the importance of them to the success of the organization's objectives.

- **Monitoring Activities:** Periodic and regular evaluation of internal control procedures is essential to ensure that internal controls are functioning effectively and as designed. Deficiencies identified have to be corrected on a timely basis and the findings reported to the appropriate authority, including top management, the board of directors, or the audit committee, as appropriate.

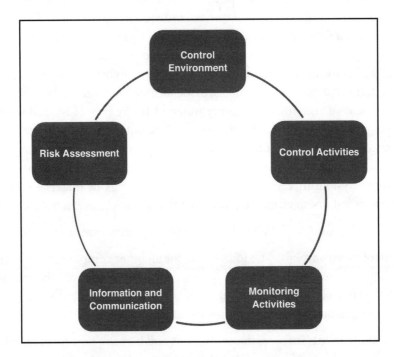

Figure 3.4 Five internal control objectives

COSO depicts the relationships between objectives, components, and operating units as a cube as illustrated in Figure 3.5. The cube clarifies that each component applies to each objective and to each unit within the organization. For example, information and communication is applicable to all three objective categories and across all units of the organization.

There is no "ideal" internal control structure that is applicable at all times and to all organizations. Instead, internal control is a dynamic structure that adapts to changes in technology, business process, and management objectives. It is expected that dissimilar entities should have a dissimilar system of internal controls. Three extraneous factors that significantly affect the appropriate design of internal controls are industry, size, and regulatory environment of the organization.

Figure 3.5   COSO framework

There are multiple layers of internal controls within an organization. These layers go from broad principle-based controls to specific functional controls. The specific functional controls cascade throughout the organization flowing from the strategy setting process and link various divisions and processes to the broader objective and governing process. There could be multiple subobjectives for each activity. For example purchasing raw materials from an outside vendor might include subobjectives of price verification, quality control, bidding process, supply chain management on use of child labor, and others.

The 2012 COSO framework sets out five components of internal control, 17 principles within each component and 81 attributes representing characteristics associated with principles as shown in Figure 3.6. The components and their respective principles are

- Control Environment
    - Demonstrates a commitment to integrity and ethical values.
    - Independence of the board of directors from management. Board of director oversight for the development and performance of internal controls.
    - Management establishes a system of effective management control.

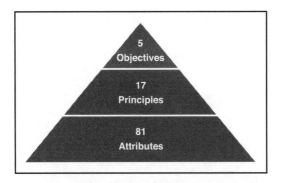

Figure 3.6 COSO Objectives, Principles, and Attributes

- A commitment to attract, develop, and retain qualified individuals.
- Holds individuals accountable for their internal control responsibilities.
■ Risk Assessment
  - Objectives are specified with sufficient clarity to enable the identification and assessment of related risks.
  - Identification, analysis, and management of risks associated with achievement of the objectives across the entity.
  - Consideration of potential fraud.
  - Change management, or identification and assessment of changes that could impact the system of internal controls.
■ Control Activities
  - Selection and development of control activities that contribute to the mitigation of risks.
  - Selection and development of control activities over technology.
  - Deployment of control activities.
■ Information and Communication
  - Obtains and generates quality, relevant information to support proper functioning of internal controls.
  - Internal communication of relevant and timely information.
  - External communication of relevant information.

- Monitoring Activities
    - Ensures proper functioning of the components of internal controls.
    - Assessment of internal control deficiencies and its timely communication to all responsible parties so that corrective actions can be undertaken.

Deficiency in internal controls refers to shortcomings in some aspect of the system that has the potential to adversely affect the ability of the entity to achieve its objectives. A deficiency in one area could be counter-balanced through compensatory controls and not affect the overall integrity of the system. There are three tiers of deficiency:

- **Non-conformity**
    - **Major:** It adversely affects the likelihood that the organization's objectives would be met.
    - **Minor:** It does not adversely affect the likelihood that the organization's objectives will be met. For instance, delay in carrying out routine maintenance of machinery.
    - **Significant deficiency:** This is less severe than a material weakness, yet is important enough to merit attention from the BOD. Multiple significant deficiencies may collectively result in a material weakness.
    - **Material weakness:** Defined as a condition in which there is a deficiency or deficiencies that there is a reasonable possibility that a material misstatement of the entity's financial statements will not be prevented, detected, or corrected in a timely manner. This requires assessing the likelihood of error and its magnitude.

The COSO framework recognizes that organizations are increasingly shifting business activities to outside service providers to economize on costs and improve efficiency. However, it should be acknowledged that outsourcing induces additional risks and creates challenges to enforce effective internal controls over an organization's operations. Clearly, although management might choose to execute activities in a manner it deems efficient and appropriate, it cannot and should not abdicate its responsibilities to monitor those activities and ensure their integrity. Consequently, the COSO framework is expected to be applied to the entire entity regardless of what choices are made in execution of business activities, either internally or through outsourcing.

The COSO framework also acknowledges innovation in technology that creates not only new opportunities but new risks. Opportunities can manifest in the development of new business markets and models or generate efficiencies. However, those innovations might also increase complexity or create new vulnerabilities, which make the identification and management of risk more difficult. Although the COSO principles do not change, the actual implementation of these principles might have to be adapted with changes in technology.

3.7 Benefits, Costs, and Limitations of Internal Controls

Internal controls provide many benefits to an organization, which include

- Added confidence to management regarding the achievement of objectives;
- Feedback on how a business is functioning, which helps prevent surprises;
- Prerequisite to access capital markets;
- Reliable and relevant information to support management decision making;
- Consistent mechanisms for processing transactions, supporting the quality of information;
- Increased efficiency within accounting function and processes;
- Ability to accurately communicate firm performance with business partners and customers.

The costs of implementing internal controls are direct, indirect, and opportunity costs. *Direct costs* are out of pocket costs that the organization incurs to hire additional employees or purchase security measures. *Indirect costs* are the costs of overhead. *Opportunity costs* are benefits that would have been attained through alternative use of resources devoted to internal controls. Specifically, these costs include

- Higher compensation cost to attract and retain qualified employees;
- Cost of procuring the latest technology;
- Efforts required to design control activities as well as to implement them;
- Maintaining an efficient channel of information dissemination.

As with every other business decision, management is expected to weigh the costs versus the benefits of alternative approaches in implementing a system of internal controls. As noted earlier, no system of internal control achieves perfection or is able to provide absolute assurance that no error is present. Hence, management has to optimize on the level of investment on internal controls and the risk it is willing to bear. Management is expected to balance the allocation of resources to the areas of greatest risk, complexity, or other factors relevant to the organization. Although the assessment of cost might be objective, the assessment of benefits is mostly subjective. The use of subjective judgment in determining the optimal extent of internal controls further increases the variability in actual controls implemented across organizations.

Inherent limitations of internal controls include

- The quality and suitability of objectives. The management, with the oversight of the Board of Directors, prioritizes the objectives and assesses risks in attaining those objectives. Internal controls are then designed to mitigate some of those risks. However, if a pertinent risk is overlooked or underestimated, requisite controls would not be designed, and the internal controls system might not be effective.

- Cost-benefit analysis. As just noted, management undertakes a cost-benefit analysis in designing appropriate controls for the organization. Because some of the costs and most of the benefits are assessed subjectively, it is possible that certain costs are over-estimated, and the corresponding benefits are underestimated. In such situations, cost-effective controls may not be put in place due to errors of estimation of costs and benefits.

- Simple errors or mistakes could lead to breakdown of the system. A system is only as good as the people running it. A perfectly designed system might fail due to human error. The importance of hiring competent and trustworthy personnel is essential to proper functioning of internal controls. In fact, many small U.S. businesses in their ICFR filings noted material weaknesses in their internal controls due to personnel issues.

- Collusion of two or more people can lead to circumvention of controls. Segregation of duties is an important element of internal controls. Many well-designed controls rely on segregation of asset handling and record-keeping duties to safeguard assets. However, if those two individuals collude, the controls are compromised, and error or embezzlements could go undetected. One such occurrence is discussed in Exhibit 3.4.

Exhibit 3.4: Embezzlement by Senior Accounting Personnel at Koss Corporation

In August 2010 the Securities and Exchange Commission charged two senior accounting professionals at Koss with accounting fraud and embezzlement of more than $30 million from the company. Koss is a Milwaukee-based headphone manufacturer with market capitalization of about $40 million in 2012. The Principal

Accounting Officer/VP of Finance embezzled millions of dollars over many years and instructed her assistant to falsify accounting books and records in an attempt to disguise theft.

Embezzlement of funds occurred through a variety of means including fraudulent cashier's checks, fraudulent wire transfers, unauthorized payments, and misuse of petty cash. The amounts embezzled over the years 2005-2010 are shown in Table 3.1. Koss's revenue, income, and assets are also shown for comparison. As is evident from the table, the embezzlement grew from about $2 million per year to more than $10 million in 2010. The total amount of funds embezzled of $30 million is 75% of the market capitalization of the company, which is significant to the size of the company.

Table 3.1   Embezzlement at Koss Corporation

	2005	2006	2007	2008	2009	2010	Total
Improper Use of Cashier's Checks	>$2 M	>$2 M	>3 M	>3.5 M	>1.3 M	>2.3 M	>14.1 M
Unauthorized Wire Transfers				>1.3 M	>7M	>7.8M	>16.1 M
Total Embezzled	>2 M	>2 M	>3 M	>4.8M	>8.3M	>10.1M	>30.2 M
Koss's Revenue	40.3 M	50.9M	46.2M	46.9M	41.7M	40.6M	265M
Koss's Earnings	4.5 M	6.2M	5.1 M	4.5 M	(0.25M)	(3.5M)	16.5M
Koss's Current Assets	23.6M	25.3M	23.8 M	24.1 M	16.4 M	15.1 M	
Koss's Total Assets	29.2M	31.4M	29.2 M	30 M	29.6M	25.75M	

The embezzlement was hidden through false entries on the company's general journal. One of the schemes was to understate sales and thereby siphon off the cash generated from those sales. The resultant impact on the income statement would be understated sales, overstated cost of sales, and understated net income.

- Management override of internal controls. In most organizations, top management is responsible for designing and communicating controls to mid-level managers and employees. Because top management is ultimately responsible for the controls, they can exercise the power and authority to override them. If a CEO or CFO were to ask a mid-level employee to make an exception to normal procedures due to extenuating circumstances, most employees would comply without further questioning. When the control is overridden, the errors it was designed to prevent could occur, leading to the failure of the internal control system. A case of override of internal controls at InfoGroup is presented in Exhibit 3.5, which discusses such an occurrence.

Exhibit 3.5: CEO's Perk and Related Party Transactions at infoUSA Inc. (InfoGroup Inc.)

In 2010, the Securities and Exchange Commission charged the CEO and Founder of infoUSA Inc. (now InfoGroup Inc.) for obtaining $9.5 million of unauthorized and undisclosed perquisites and for undisclosed related party transactions of approximately $9.3 million. Over a period of five years (2003 to 2007) the CEO's personal expenses that were reimbursed by the company included private jet travel, personal travel, club memberships, more than 20 cars, and premiums for personal life insurance. The details are included in Table 3.2. The CEO had exempted himself from the company's reimbursement policies and hence was able to charge the company for his personal expenses.

Table 3.2 CEO Perks at infoUSA Inc. (InfoGroup Inc.)

Description	2003	2004	2005	2006	2007	Total
Reported Perquisites	$6,000	$6,000	$7,000	$113,000	$818,000	$950,000
Actual Perquisites	$2,169,000	$2,625,000	$1,968,000	$1,740,000	$1,916,000	$10,418,000
Undisclosed Perquisites	$2,163,000	$2,619,000	$1,961,000	$1,627,000	$1,098,000	$9,468,000

(Source: SEC Complaint: http://www.sec.gov/litigation/complaints/2010/comp21451-gupta.pdf)

The CEO also held primary ownership in two additional companies, which for part of the time operated out of the facilities of infoUSA Inc. Specifically, infoUSA Inc. paid the two companies owned by the CEO for the lease of his aircraft and costs for his homes, yacht, and automobiles. In addition it provided office space to these companies without charging rent. The specifics of related party transactions are shown in Table 3.3.

Table 3.3 Related Party Transactions at infoUSA Inc. aka InfoGroup Inc.

Description	2003	2004	2005	2006	2007	Total
Aircraft (paid to Annapurna)	$1,762,000	$929,000	$265,000	$ -	$ -	$2,956,000
Aircraft (paid to jet leasing company)	$1,346,000	$836,000	$179,000	$ -	$ -	$2,361,000
Homes (paid to Annapurna)	$100,000	$120,000	$ -	$ -	$ -	$220,000
Yacht (paid to Annapurna)	$370,000	$473,000	$32,000	$ -	$ -	$875,000
Autos (paid to Aspen Leasing)	$41,000	$58,000	$ -	$ -	$ -	$99,000
Rent free use of infoUSA's building by Gupta's Entities	$19,000	$19,000	$ -	$ -	$ -	$38,000
Total Amount of Leases with Gupta's Entities	$3,638,000	$2,435,000	$476,000	$ -	$ -	$6,549,000
Aircraft Purchased from Annapurna (not disclosed)	$1,099,000	$1,650,000	$ -	$ -	$ -	$2,749,000
Autos Purchased from Aspen Leasing (not disclosed)	$ -	$ -	$182,000	$ -	$ -	$182,000
Quest Ventures (not disclosed)	$260,000	$ 307,000	$475,000	$995,000	$1,831,000	$3,868,000
Total Undisclosed Related Party Transactions	$2,765,000	$2,870,000	$836,000	$995,000	$1,831,000	$9,297,000

(Source: SEC Complaint: http://www.sec.gov/litigation/complaints/2010/comp21451-gupta.pdf)

3.8 Incorporation of Fraud Risk in the Design of Internal Controls

The COSO framework explicitly states that management's risk assessment in the design of controls should consider risks related to fraudulent reporting as shown in Figure 3.7. The revised framework enhances consideration of anti-fraud expectations and includes consideration of the potential for fraud as a principle of internal control. Further, in response to the Foreign Corrupt Practices Act in the U.S.,[13] it also requires management to consider possible acts of corruption by its personnel as well as by external parties, which impairs its ability to comply with laws and regulations in their respective jurisdiction. It requires management to assess risks if the entity's standards on ethical conduct are not adhered to by management, other personnel, and outsourced service providers. It is important to note that the assessment of risks has been extended outside the organization's boundary to include outsourced service providers. This in a way holds management responsible for some actions of its suppliers and effectively prevents management from outsourcing corruption. That is, some U.S. business operating in developing nations with rampant corruption might engage a third party as consultants who essentially bribes officials in that country; engaging such parties would violate this principle. Paragraph 264 of the COSO framework states that management can stipulate the expected level of performance standard through contractual relations and maintain oversight of third-party actions.

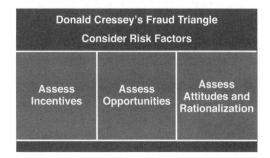

Figure 3.7   Assessing risk of fraud

Material misstatements in financial reporting could be a result of willful omission or misstatement. These events occur through unauthorized receipts, expenditures, financial misconduct, or other irregularities. Regardless of who commits the fraud, the responsibility and accountability for anti-fraud policies and procedures rests with management. Hence, as part of risk assessment and ensuing design of controls, COSO requires consideration of the following:

- Degree of estimates and judgments in external reporting
- Fraud schemes and scenarios common to the industry sectors and markets in which the entity operates
- Incentives that may motivate fraudulent behavior
- Nature of automation
- Unusual or complex transactions subject to significant management influence

- Vulnerability to management override and potential schemes to circumvent existing control activities.

When fraudulent activity is detected, remediation could be necessary. It might not be sufficient to deal with the instance and person(s) involved in isolation, but it might require rethinking the design of internal controls to prevent such occurrences in the future.

3.9 Legislation on Internal Controls

To prevent a repeat of the scandals that cost investors billions, companies are now required in annual reports to have chief executives and chief financial officers attest to "the effectiveness of the internal control structure and procedures of the issuer for financial reporting." The outside auditors also have to test the internal controls and give an opinion on their effectiveness.

The initial costs of developing the required internal controls can run as high as $5 million to $10 million for a small company once it goes public because private firms do not have to have the same accounting and compliance structures in place as larger corporations. In addition to the expense of putting in place the program for the first time, the annual review by a company's accountants adds to the costs of the audit.

An analysis of all reported material weaknesses in 2010 revealed that a higher proportion of material weakness was reported by smaller firms based on total revenue or assets. Smaller firms tended to have more internal control material weaknesses due to improper segregation of duties, lack of competent accounting personnel, and lack of independent audit committee. Although the large companies, measured in terms of assets or revenue, tended to report less material weaknesses in internal control, when they did it was related to complex accounting transactions or information technology.

3.10 Chapter Summary

This chapter has covered many important concepts critical to the proper functioning of modern organizations, worldwide. Foremost, the theoretical concepts that form the foundation for corporate governance discussions and policies were discussed. A widely acceptable economic theory, known as the agency theory, provides a tractable way to analyze the conflicts of interest that can exist between various parties in a modern organization. There are two widely studied ramifications of this theory: moral hazard and adverse selection. Moral hazard, which hypothesizes suboptimal behavior in absence of commensurate negative consequences, is of particular significance to the discussion of corporate governance. Corporate governance mechanisms are developed to counter the perceived issues caused by moral hazard. The issues, however, may vary across different cultures and regulatory regimes based on applicable laws, cultural norms, and acceptable business practices. Hence, an understanding of these differences is imperative in prescribing or evaluating corporate governance measures across nations. In Western developed economies, where ownership is dispersed, principal-agent conflicts are prevalent. However, in developing economies with strong family businesses and ties, the principal-principal conflict tends to be more relevant.

With increased globalization, which has brought forth not only global expansion, but also increased competition, the balancing of organizational flexibility with control has become challenging. Flexibility in an organization is important to ensure profitable opportunities are exploited on a timely basis. However, enforcing control across wide geographical areas and diverse cultures requires imposing limitations on behavior and curtailing flexibility. The four levers of control present a potential solution to this dilemma. They recommend establishing corporate objectives and culture to communicate desirable actions, setting up boundary conditions to prevent undesirable actions or decisions, and creating a measurement system for continuous monitoring of key organizational metrics.

A properly designed system of internal controls is essential in modern organizations to ensure that the objectives of the organization are being achieved in an efficient manner. Though there is no "silver bullet" design of internal controls, a properly functioning internal control system reduces the possibility of material errors or deviations from organizational objectives. The COSO framework serves as a de-facto benchmark for internal control design. It outlines five objectives, which are further expanded to 17 principles and 81 attributes. The COSO framework provides broad guidelines on the design of adequate controls across organizations. However, adequacy of controls is compromised by management's limitation to comprehensively assess all risks pertinent to the organization and the proficiency/competency of its employees. The optimal design of controls is additionally hampered by the subjectivity inherent in assessing costs and benefits. In summary, designs of appropriate internal controls vary across organizations and change over time; hence, periodic assessment of the adequacy of controls is highly recommended for all and required for some.

3.11 Endnotes

1. Two seminal works in agency theory are M. Jensen and W. Meckling, "Theory of the Firm: Managerial Behavior, Agency Costs and Capital Structure," from *Journal of Financial Economics* and Eugene F. Fama, "Agency Problems and the Theory of the Firm," from *Journal of Political Economy*.

2. Mark S. Beasley, An Empirical Analysis of the Relation between the Board of Director Composition and Financial Statement Fraud, 71 *The Accounting Review* (1996). Patricia M. Dechow, Richard G. Sloan and Amy Sweeney, Causes and Consequences of Earnings Manipulation: An Analysis of Firms Subject to Enforcement Actions by the SEC, 13 *Contemporary Accounting Research* (1996). April Klein, Audit Committee, Board of Director Characteristics, and Earnings Management, 33 *Journal of Accounting and Economics* (2002).

3. See definition in R.E. Hoskisson, L. Eden, C.M. Lau and M. Wright, "Strategy in Emerging Economies," *Academy of Management Journal*.

4. Michael Young, M.W. Peng, D. Ahlstrom, G.D. Bruton and Y. Jiang, "Corporate Governance in Emerging Economies: A Review of the Principal-Principal Perspective," from *Journal of Management Studies*.

5. M.W. Peng and P. Heath, "The Growth of the Firm in Planned Economies in Transition: Institutions, Organizations and Strategic Choice," from *Academy of Management Review*.

6. An Australian journalist, Hamish McDonald, has a book on the Ambanis titled *The Polyester Prince* followed by *Mahabharata in Polyester*.

7. "Fuelling the Feud," from *The Economist*, http://www.economist.com/node/16082277.

8. M. Faccio and L. Lang, "Young Dividends and Expropriation," *American Economic Review*.

9. M. Backman, *Asian Eclipse: Exposing the Dark Side of Business in Asia*.

10. For instance, break-up of Reliance Industries in India due to sibling rivalry.

11. L.R. Bautista, "Ensuring Good Corporate Governance in Asia," from *Recreating Asia: Visions for a New Century*.

12. Dutta, S.K., D. Caplan and D. Marcinko (2013), Growing Pains at Groupon, *Issues in Accounting Education*

13. And a similar and more restrictive Act in the U.K.

4

Detection of Fraud: Shared Responsibility

Like a torrent of cold water the wave of publicity raised by the ... case has shocked the accountancy profession into breathlessness. Accustomed to relative obscurity in the public prints, accountants have been startled to find their procedures, their principles, and their professional standards the subject of sensational and generally unsympathetic headlines.
—Editorial, *Journal of Accountancy*

4.1 Introduction

Although the editorial you just read could be applicable in many recent occasions—the financial crisis of 2008, the corporate scandals of Enron, WorldCom in 2002, or Savings and Loan crisis of the 1980s, this editorial is actually from 1939 in the aftermath of corporate scandals of the time, Kreuger and Toll, Inc. and McKesson and Robbins. These two cases of the earliest corporate fraud are discussed in Exhibit 4.1. Public outcry regarding investor losses arising from these corporate failings led to the requirement of mandatory audits for public companies.

Exhibit 4.1: Earliest Instance of Financial Fraud

Kreuger and Toll (1932)

Kreuger and Toll was a Swedish company that specialized in matches and had cornered about 75% of the European and American markets. To obtain exclusive rights to sell matches in certain countries, the company offered significant loans to various European governments ravaged by World War I. It turned to the American financial markets to raise capital and attracted investors by offering them up to 20% annual dividends. The bold plan ran into trouble when several countries began defaulting on their loans. To maintain the charade, the company resorted to a Ponzi scheme, paying dividends to existing investors from capital raised from the new investors. Thus, they became dependent on continually obtaining funds from new investors to pay their increasing obligations. Finally, when the funds from new investors dried up during the Great Depression, the scheme collapsed, and the company ran out of funds to pay its dividends to shareholders and interest and principal to the lenders. The fraud at Kreuger and Toll built the support for the passage of the U.S. Securities Act of 1933 and 1934 and led to the creation of the Securities and Exchange Commission (SEC) to oversee corporate financial reporting.[1]

McKesson and Robbins (1938)

The fraud at McKesson and Robbins was orchestrated by a career criminal who gained control of the company by using an alias. Once in control, he placed his brothers in key position within and outside the company. One of his brothers opened up a fictitious company, W.W. Smith and Company, which would write purchase orders bearing the names of fictitious companies and mail those to McKesson and Robbins. Another brother was in charge of the shipping department where he would forge documents to give the appearance that the products had been shipped to legitimate customers. His fourth brother was appointed as the company's treasurer who had control of cash and could create the appearance that cash was being received from customers and disbursed to suppliers. Their profit was the commission that W.W. Smith received per sale. The fraud was finally discovered in 1938 by an employee who got suspicious of W.W. Smith and undertook an investigation, which was followed by an SEC investigation.

After examination and analysis of what enabled company CEOs to commit a fraud over an extended period of time and remain undetected, the SEC recommended that non-officer members of the Board of Directors nominate the auditors. This, in some sense, was the precursor of the requirement and responsibility of Audit Committees within the Board of Directors. Also, the American Institute of Certified Public Accountants (AICPA) made observing inventory and confirming accounts receivables standard audit procedures.

In the post-mortem of the recent financial crisis in the U.S., the PCAOB's Investor Advisory Group made a presentation on March, 2011 titled, "The Watchdog that Didn't Bark …Again," a critical review of the audit profession. It asserted that the auditing profession failed the test in the crisis as dozens of the world's leading financial institutions failed when each one of them had received a "clean bill of health" from the Big 4 auditors just months before their demise. Detailed information on U.S. financial firms that either failed or had to be rescued is shown in Table 4.1. As a result, serious questions are being raised both about the quality of the financial reporting practices at these firms and the quality of audits that permitted those reporting practices to go unchecked. Although the auditors did not cause the financial crisis, they failed to sound the warning on the looming crisis. They did not necessarily design complex transactions that obfuscated the true financial health of the companies, but they permitted their use. These practices, it is argued, deprived investors of important and timely information and might have encouraged companies to take on risks. The U.K.'s FSA has labeled it as "a worrying lack of professional skepticism."

Table 4.1   Financial Firm Failures During the 2008 Financial Crisis

Company	Event	Date	Investor Losses ($m)	Audit Firm
Citigroup	TARP	10/28/2008	212,065	KPMG
AIG	TARP	9/16/2008	156,500	PwC
Fannie Mae	Gov't Takeover	9/6/2008	64,100	Deloitte
Freddie Mac	Gov't Takeover	9/2/2008	41,200	PwC
Lehman	Bankruptcy	9/15/2008	31,437	E&Y
Washington Mutual	Bankruptcy	9/26/2008	30,559	Deloitte
Countrywide	Fire-Sale	1/11/2008	22,776	KPMG
Bear Stearns	Fire-Sale	3/17/2008	20,897	Deloitte
New Century	Bankruptcy	4/2/2007	2,576	KPMG

All parties involved throughout the financial reporting process bear some responsibility in preventing and detecting fraud; however, a disproportionate share of blame tends to be borne by the external auditors. The auditing profession has often argued that this could be due to the perceived "deep pockets" of the auditing firms. Investors and creditors might expect the auditing firms to act in an insurance capacity and reimburse them for their losses. In the wake of financial reporting problems, it is often forgotten that ensuring the validity of the financial report is a collective responsibility. The various parties, which include the management, Board of Directors, audit committee, internal auditor, and external auditors, have complementary and overlapping responsibility to maintain the integrity of the financial reporting process. They are, individually and collectively, responsible for not only delivering a high quality financial statement but also to deter and detect fraud. The relationship between these parties and a summary of their roles in preventing fraud is pictorially illustrated in Figure 4.1.

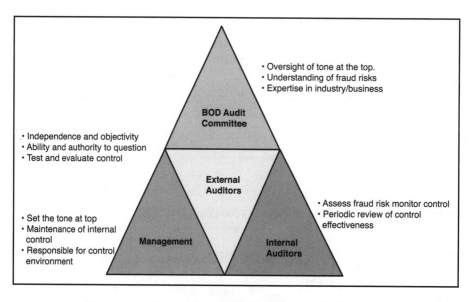

Figure 4.1 Financial reporting process: A collective effort

This chapter first discusses the expectations gap that confronts the accounting/auditing profession, worldwide. Next, the responsibilities of all involved parties in prevention and detection of fraud or financial misstatements are identified and analyzed. External auditors in their responsibility as a third-party observer of the financial reporting process bear a disproportionate responsibility on the failings of an organization's financial reporting system. The Board of Directors (BOD) and audit committee are in an oversight role, and their responsibility includes ensuring that the organization functions under proper control environment. Management has the primary responsibility for the financial reporting process and for implementing controls. Internal auditors play a key role in a company's internal control structure and are entrusted with the responsibility to assess the effectiveness of the internal control system and periodically assess fraud risk. The chapter concludes by summarizing empirical evidence on parties responsible for detecting financial fraud in the U.S.

4.2 Expectations Gap in the Accounting Profession

The *expectations gap* is the difference in attributes between what is expected by customers and what is provided by the supplier. In an auditing context, it is the difference between what the users of financial statements perceive the role of an audit to be and what the auditing profession claims is expected of them. There is a widespread perception among the public and regulators[2] that audits are not fully accomplishing their objectives. Specifically, the public's perception is that audits ensure the quality of the financial statement and are a testament to the financial health of the company; however, the audit process only assures that the financial statements are free from material errors and is not a value judgment on the health or future prospect of the company. Many government appointed committees, primarily in the U.S., U.K. and Canada,[3] have studied this issue and have recommended steps to reduce the gap.

Empirical research evidence indicates that the expectation gap between the auditors and the investing public is a world-wide phenomenon. The gap has been shown to exist in the U.S. in areas of independence, going concern, fraud, and illegal acts.[4] In addition to U.S. and U.K. this gap is also documented in countries under diverse regulatory and cultural settings including Australia,[5] Bangladesh,[6] Canada,[7] Egypt,[8] Iran,[9] Malaysia,[10] Nigeria,[11] Singapore,[12] and others. Although there are some suggestions that the gap could have resulted from lack of technical competence of auditors or their low commitment to public interest,[13] the reason most often espoused is that societies judge auditor performance based on outcomes and expect them to perform duties beyond those prescribed or can be reasonably expected of them.[14] An academic study explored the issue further and determined that the expectations gap was subdivided into three components: substandard performance (16%), deficient standards (50%), and unreasonable expectations (34%).[15] As deficient standards are relatively easy to fix through a process of standard settings, it is the component that is most easily addressed. Recent advancement in this area has come through the formation of PCAOB in the U.S., the FRC in U.K. and the IAASB internationally.

In October 2010, U.S. Securities and Exchange Commission sponsored a panel discussion titled "Responsibility for Preventing and Detecting Financial Reporting Fraud."[16] Ms. Cynthia Fornelli of the Center of Audit Quality summarized this shared responsibility as

> "First and foremost, of course, is management. They bear the primary responsibility for deterring and detecting fraud within an organization. So they are responsible for internal controls, for having a robust ethical program, for establishing and maintaining the whistle-blower program. Then of course are the boards of directors and the audit committees. They have the responsibility for overseeing the business and control environment within an organization. Then in our minds came internal audit, if indeed the corporation has that function and of course a listed company is required to have an internal audit function. They don't necessarily have to have an internal audit department but the internal auditor function plays a really key role in helping the company create the internal control structure within an organization. And then last but not least are the external auditors who

must be independent of the company and who provide a public report on the entity's annual financial statements, including for public companies over a certain market cap an assessment of the effectiveness of an entity's internal controls over financial reporting."

4.3 Responsibility of the External Auditor

External auditors express an opinion that the financial statements are fairly stated in all material aspects. The regulators across various jurisdictions, in an effort to protect investors, require that financial statements of publicly traded companies be audited by independent auditors to ensure compliance with the accounting practices (GAAP or IFRS) in their jurisdiction. Thus, the function of external audit adds credibility to the financial reports. Additionally in the U.S., post Sarbanes-Oxley Act, auditors are also required to opine on the adequacy of the internal controls over financial reporting.

While the auditing profession in the late 1990s tried to deflect additional responsibility for fraud detection, subsequent standard settings in the U.S. (SAS 53, SAS 82 and SAS 99) have progressively increased auditor's responsibility in detecting fraud. In 2006, the ASA 240 was issued in Australia, which increased the external auditor's responsibility in the area of fraud detection.[17] In the U.S., a study by the Treasury Department in 2008 recommended that the audit report explain the auditors' role and their limitations in detecting fraud.[18] In a similar vein, a current proposal at the European Union requires auditors to disclose whether the audit was designed to detect fraud.[19] The U.S. Treasury Department study noted the following:

> Public investors have appropriately raised questions when large frauds have gone undetected. Among the attributes that the public expects of auditors is a clear acknowledgement of their responsibility for the reliability of financial statements, particularly with respect to the detection of fraud, notwithstanding the recognition that a company's management and board have the primary role in preventing fraud. The public may believe that auditors will detect more fraud than those in the profession believe can be reasonably expected.... There are difficulties of detecting fraud, especially before it has resulted in material misstatement....

Authoritative guidance in the U.S. has directly addressed auditor responsibility in detecting fraud. Statement of Auditing Standards No. 99, issued by the American Institute of the Certified Public Accountants (AICPA), requires auditors to obtain information to identify financial statement fraud risk and take that risk into account while designing audit procedures. It also provides guidance in responding to the results of the risk assessment and documentation of the auditor's risk assessment. Public Company Audit

Oversight Board (PCAOB) in its Section 110, Responsibilities and Functions of the Independent Auditor, paragraph .02 states, "The auditor has responsibility to plan and perform audit to obtain reasonable assurance about whether the financial statements are free of material mis-statements, whether caused by error or fraud." This indicates that auditors bear some responsibility for detecting fraud. That is, if employee fraud or corporate fraud leads the financial statements to be materially mis-stated, it is the auditor's responsibility to detect such causes. PCAOB's Auditing Standard No. 5 (AS 5) titled, *An Audit of Internal Control over Financial Reporting that is Integrated with an Audit of Financial Statements*, increases emphasis on assessing fraud risk and antifraud controls. The standard outlines audit procedures to test the effectiveness of anti-fraud controls in preventing and detecting financial statement fraud.

External auditors are in a unique position to detect financial statement fraud due to their financial expertise, access to all necessary financial information, and authority given to them by the BODs. It is recommended in professional standards worldwide that auditors conduct their work with a degree of professional skepticism and pay special heed to fraud symptoms or red flags that can arise during the course of their work. The auditors should also diligently follow through with employee allegations of financial impropriety and whistle-blower allegations. In such an instance, it is recommended that the auditors share the information with the audit committee and other appropriate parties. Sometimes, sharing of seemingly trivial information may uncover well-woven fraud schemes.

Additionally, the auditors might employ a risk-based approach to designing the audit procedure, which leads to frequent testing of high-risk areas, making it more likely to detect fraud should it exist. With increasing complexity of modern businesses and their use of complex accounting transactions, it is becoming increasingly important for auditors to become familiar with their client's businesses. This includes having sufficient knowledge of the industry and operating strategies. The knowledge of these would enable the auditor to independently assess the risks and complexities in the business and their ramification on financial reporting. Proper audit procedures can then be designed to obtain assurance in these areas. A CFA (Certified Financial Analyst) Institute survey conducted in 2008 indicated that it was important to have auditors identify key risk areas, significant changes in risk exposures, and amounts either involving high degree of uncertainty in measurement and significant assumptions or requiring a higher level of professional judgment.[20]

Auditing standards from the PCAOB caution the external auditor that management has a unique ability to perpetrate fraud because it is in a position to directly manipulate accounting records and present fraudulent financial information. This often involves management override of controls that otherwise would appear to operate effectively. In fraud, unlike unintentional errors, there is an attempt to conceal. Concealment is undertaken through either withholding information or through its willful misrepresentation. It may also involve collusion among management, employees, or third parties. Fraud is more difficult to detect because auditors may be unknowingly relying on fraudulent evidence that appears to be valid. Furthermore, audit procedures that are effective in detecting errors may not be effective in detecting fraud. Hence, the auditor should conduct the audit with a mindset that recognizes the possibility that fraud may be present, notwithstanding its prior experience with the company and its management.

External auditors are expected to exercise professional skepticism, which is "an attitude that includes a questioning mind and a critical assessment of audit evidence." It implies evaluating and challenging audit evidence and allowing for the possibility that the evidence is deliberately false. It also means remaining alert for information and following up on various leads. While the auditor works closely with top management, which has a similar objective in ensuring that the financials are free from material error, the auditors should consider the possibility that management might be involved in fraud, hence the auditors should skeptically examine override of controls by the management and seek justification. Through this level of scrutiny, auditors not only increase the likelihood that fraud is detected but also the *perception* that it will be detected, thereby acting as a deterrent.

4.4 Responsibility of the Board of Directors

The separation of ownership and control in modern organizations, discussed in detail in Chapter 3, "Preventive Measures: Corporate Governance and Internal Controls," requires the Board of Director (BOD or Board) to play a crucial role in the proper functioning of the system. The BOD lessens the impact of principal-agent conflict of interest by being the owner's "eyes and ears" within the organization. The Board is also involved in critical financing decisions of the organization and in that function safeguards invested capital and the interests of the current owners against over-dilution of their interest/control. The board assesses management performance and approves the compensation package for the management. The Board is granted an oversight function over management, which includes hiring, firing, and evaluation of managerial performance. The Board of Directors is also expected to function in the best interest of the company and its shareholders. Not only does it serve in an oversight function, but it also serves as an important resource to the management in that it is expected to provide consultation and advice.

The effectiveness of the oversight function depends on the Board's independence, competency, authority, and resources. An independent board is more effective in its oversight function and can objectively assess and compensate managers for their performance without fear of reprisal. On the other hand, if the management exercised significant control over the Board through its power to nominate or remove Board members, the Board compromises its independence and effectively becomes management's "rubber stamp." Some academic research suggests that Board's independence is affected when the CEO also serves as the Chairman of the Board.

Competency of Board members is also an important characteristic. Without competency and knowledge the Board might be unable to evaluate management decision. Although lack of independence compromises a Board's effectiveness, lack of competency of Board members renders the Board useless. The Board should have adequate resources available to them to carry out their function. For a Board to be effective it has to meet at frequent intervals and be constantly informed of major corporate initiatives, opportunities, risks, and challenges. Although having independent Board members gives the appearance of being effective, for the Board to be truly effective it has to be comprised of competent individuals with adequate resources and authority available to them.

An effective Board through its vigilant oversight can significantly reduce the likelihood of certain types of fraud occurring, especially management fraud or issuance of fraudulent financial statements. The Board, through its audit committee, usually oversees the financial reporting process and the design of internal controls. As accounting rules and internal control design requires specific technical knowledge, it is a requirement in the U.S. that all members of the audit committee should possess financial literacy and at least one should have financial training. A competent and vigilant Board could also prevent management override of internal controls. As discussed in the previous chapter, one of the serious limitations of internal controls was the possibility of management override of control. The Board could institute a communication mechanism whereby it is alerted whenever a key control is overridden and evaluates whether the exception was justified.

The effectiveness of the Board can be enhanced by creating open channels of communication with various "key" parts of the organization. If the bylaws of the organization restrict access to the Board and of the Board, it severely limits its effectiveness. If the Board only receives information that is "filtered" by the top management, it compromises the Board's ability to properly evaluate the management and may be oblivious to fraudulent actions committed by the management. At a minimum, the Board should have direct access to the Internal Audit Group and vice versa. Similarly, the Board should be actively engaged in selection and retention of external auditors. The Board should institute an open access policy to all employees and protect whistle-blowers from management reprisals.

In summary, a Board's effectiveness in fraud prevention could be enhanced through the following:

- With independence and competence of the Board and its individual members
- By addressing and discussing fraud risks and possibilities on a regular basis
- By maintaining an open communication channel with various parts and levels within the organization, especially financial and operational functions and most importantly with the internal audit group
- By being proactive in decisions regarding selection and retention of external auditors; meeting with external auditors to discuss their findings on a periodic and regular basis
- By being vigilant of management actions and periodically assessing risks and reviewing risk abatement policies
- By periodically reviewing fraud prevention and detection policies of the management
- By setting up proper procedures of resolving allegations of financial misappropriation within the organization; ensuring thorough investigation and demanding a full report of the resolution

As the Board is entrusted with intimate knowledge about a company's strategy and functioning, the Board members are expected to maintain a high level of personal ethical standard. The Board members are usually highly accomplished individuals, have had illustrious business careers and are independently wealthy. Although they are privy to much confidential and valuable information, they are expected to adhere to the highest level of ethical standards and not violate the trust. In Exhibit 4.2, one such violation of trust by a former Goldman Sachs Director and its ramifications are presented.

Exhibit 4.2: Trading on Insider Information

In October 2012 a former Director of the investment bank Goldman Sachs and the consumer product company Proctor and Gamble was sentenced to two years in prison for providing insider information to a hedge fund manager. The hedge fund manager who ran Galleon Management was also convicted and is serving an eleven year prison sentence.

The case was built by matching phone calls exchanged between the Director and the hedge fund manager from 2007 to 2009 and subsequent trading done by the hedge fund on the shares of Goldman Sachs and Proctor & Gamble. The prosecution alleged that as a result of these trades the hedge fund netted $16 million through gains or avoidance of losses.

One of the most persuasive pieces of evidence for prosecution was based on the Warren Buffet Goldman Sachs arrangement struck soon after the collapse of Lehman Brothers in 2008. Records of phone calls were produced that showed that at 3:13 p.m. on September 23, 2008, the Director had called into a Goldman Sachs boardroom meeting in which Warren Buffet's investment of $5 billion into Goldman Sachs was discussed. At 3.54 p.m. the Director placed a call to the hedge fund manager. The content of the call was not available. However, soon thereafter at 3.56 p.m. the hedge fund purchased $43 million of Goldman stock. These shares were sold the following day, after the news of Warren Buffet's investment in Goldman Sachs was made public, netting a profit of $1.2 million for the hedge fund in less than a day.

This Director had joined the Goldman Sachs Board in November 2006 amidst high praise from the CEO for his "strategic and operational expertise and judgment." Later, in testifying about the importance of the September 23, 2008, board meeting, he emphasized that board members are not authorized to disclose any information from any board meeting. Further, the CEO testified, that doing so is a betrayal of fiduciary duty to shareholders.

4.5 Role of the Audit Committee

Since the late twentieth century, many companies have formed audit committees comprising of outside directors as a standing committee of the Board of Directors. They were delegated the task of interacting with the external auditors and providing general oversight of the financial reporting process. Since the passage of SOX in 2002, new rules were set in the U.S. by the Securities and Exchange Commission (SEC) and the various stock exchanges that mandated audit committees for publicly traded companies and provided guidelines on their composition and responsibilities. In this section, the underlying principles of the audit committee requirements are discussed.

The desired characteristics and the importance of audit committees to financial reporting are succinctly stated by Arthur Levitt, the former Chairman of the SEC. He made the following statement in a speech in 1999, a couple of years prior to the problems at Enron and WorldCom and three years prior to SOX.

Effective oversight of the financial reporting process depends to a very large extent on strong audit committees. Qualified, committed, independent and tough-minded audit committees represent the most reliable guardians of the public interest.[21]

An audit committee is usually comprised of three to five outside directors who are independent from the management. SOX specifically requires that the audit committee be composed entirely of independent members. Financial literacy of the audit committee members is a must, and financial expertise of some of the members is desirable. The requirements in the U.K. and the European Union are similar to those in the U.S. Table 4.2 provides comparative audit committee guidance in the U.S., U.K., EU, and Japan.

Table 4.2 A Summary of International Guidelines on Audit Committees

Requirements	U.S. SOX or SEC	UK FRC	EU[22]	Japan[23]
Composition	At least three members Independent	At least three members Independent	Three to five members and at least one independent member	At least three, majority should be independent
Skills of Members	Financial literacy for all At least one financial expert	Recent financial experience	At least one member with competence in accounting Financial literacy of all members	
Resources	Funds to engage outside advisors External auditors	Access to the services of employees Access to information in timely manner Funds available to seek expert advice	Access to services of employees Funds to engage outside advisors	
Oversight of External Auditors	Hiring of external auditors Oversight of work plan Approve fees and nonaudit engagements	Recommendation on hiring Annual assessment Approve fees Review nonaudit services Investigate resignation	Review audit plan Evaluate sufficiency of business risks identified by external auditor Review findings	Assess audits Appoint or discharge auditors

Requirements	U.S. SOX or SEC	UK FRC	EU[22]	Japan[23]
Oversight of Internal Auditors	Oversight of internal audit Direct channel to Internal Audit Group	Monitor and review effectiveness Approve hiring/firing of head of internal audit	Annual review of effectiveness	Ensure introduction and proper functioning
Oversight of Internal Controls	Review ICFR Understand the procedures for assessing effectiveness	Receive report from management on effectiveness of internal controls Review and approve statement in the annual report related to internal control	Monitor effectiveness of controls	Evaluate and make improvements Evaluate CEO's policies for strengthening controls
Handling of Complaints	Ensure confidentiality and anonymity of whistle-blower Protocol for receiving and investigating complaints	Review arrangements for confidentiality Ensure proper arrangements for investigation and follow-up actions	Ensure arrangements for investigation and follow-up actions Review confidentiality arrangements	
Communication with Shareholder		Separate section in the annual report		CEO prepares annual report on internal audit and control
Other			Assess appropriateness of accounting policies, judgments, and estimates Treatment of complex transactions	

To assert the importance of an audit committee and to ensure that it is given the appropriate authority, the Financial Reporting Council of the U.K. in its Guidance to Audit Committee, provides for a separate section in the annual report that describes the work of the committee. Provisions, such as these, put the spotlight on the Audit Committee and enhance its perceived importance and make it commensurate with its importance in preserving the integrity of the financial reporting process.

In some jurisdictions, there are no prescribed duties for the audit committee. Financial Reporting Council of the U.K. acknowledges that audit committee arrangements need to be proportionate to the task and will vary according to the size, complexity, and risk profile of the company. However, the audit committee is generally expected to interact with other parties in the financial reporting supply chain, which includes management, accounting personnel, internal audit group, external auditors, and the board of directors. The audit committee is also expected to set up procedures and protocols to handle allegations of financial misconduct.

In describing the functions and role of the audit committee, the verbs commonly used are "oversight," "assessment," and "review," which signifies the monitoring function of the committee. It is not expected or required to actually carry out the functions that are the responsibility of other parties, such as management's responsibility to prepare financial statements and external auditors' responsibility to plan the audit. Sometimes however, the oversight function may lead to detailed work, especially when investigating allegations on management misconduct or management overrides of controls.

The audit committee is expected to oversee the adequacy and effectiveness of the company's internal control system. Section 404 of SOX has made periodic review of internal controls even more important. Consequently, the audit committee is expected to review the management's annual report on internal controls over financial reporting (ICFRs) as well as the external auditor's assessment of management report. The audit committee should follow up on the areas that have been identified as "significant deficiencies" or "material weaknesses" and require that corrective measures are implemented on a timely basis. If the material weaknesses and significant deficiencies have not been remedied, the audit committee should assess the impact of those on the integrity of financial reporting. In cases where the impact is significant it should bring it to the attention of the entire Board and consider making additional disclosures to appropriate parties.

The audit committee is also expected to oversee the financial reporting process and review the financial statements. They should specifically assess the appropriateness and reasonableness of accounting principles used, the various estimates made, and the reserves accounted for. In the U.S., the committee is required to meet annually with the external auditors and review their findings and recommendations. In fact, a new proposal on Audit Reform by the European Commission requires auditors to prepare a longer and more detailed report for the audit committee that includes a description of the audit work carried out.

In summary, the audit committee serves a critical oversight role in ensuring the integrity of the financial reporting process. Due to its authority to hire/fire external auditors, appoint the head of the internal audit group, and oversee management functions related to financial reporting, it carries significant burden and responsibility toward ensuring that financial statements are free from material errors or fraud. While the audit committee is not engaged or responsible for the detailed work and review, it performs a high-level review of the entire process. The Deloitte Forensic Center has suggested ten useful tips for audit committees.[24] These are

1. Assess the risk of fraud by management. This risk is based on management's pressure to meet earnings expectations.

2. Evaluate internal controls that address each of the fraud risks.

3. Evaluate internal auditors, testing of the controls related to the fraud risks.

4. Is the company's management creating and maintaining an appropriate "tone at the top" and whether their actions and behavior are in harmony with their words.

5. Evaluate changes in accounting principles and estimate and question the appropriateness as well as purpose for those changes.

6. Ensure that the internal auditors have direct access to the audit committee.

7. Ensure that the internal auditors have sufficient resources and authority to conduct periodic assessment of controls and fraud detection tests.

8. Review each quarterly and annual financial statement to ensure that the risks and objectives of the organization are properly outlined in the MD&A and the assertions made in the MD&A are adequately supported by the financial statements.

9. Use current technology for fraud detection to isolate suspicious transactions on a "real-time" basis.

10. Insist that the external audit team has at least one fraud specialist to critically evaluate these items.

4.6 Management's Role and Responsibilities in the Financial Reporting Process

The primary responsibility for the quality, integrity, and reliability of the financial reporting process lies with the management of the organization. All regulator and standards, worldwide, unequivocally place this responsibility on the management as it has complete operative control and authority over the functioning of the organization. The 1933 Securities Exchange Act in the U.S. requires management to provide investors with financial and other important information relevant to the valuation of securities available for public sale. The Foreign Corrupt Practices Act (FCPA) imposed on the management a requirement to maintain functional internal controls.

The Sarbanes-Oxley Act of 2002 further increased management responsibilities over financial reporting. It made senior executives, primarily the CEO and CFO, directly responsible for maintaining adequate internal controls over financial reporting. It required management to certify the company's internal controls. The U.S. Securities and Exchange Commission implemented the legislation to specifically require the CEO and CFO of all publicly traded companies in the U.S. to certify that

- The organization has designed appropriate controls
- They provide reasonable assurance on the fairness of the financial statements
- Effectiveness of those controls are evaluated and that the controls are found to be effective
- They have disclosed any changes made to the control structure that could affect financial reporting and reasons thereof.

Additionally Section 404 of SOX requires management to document and assess the design and operation of the company's internal controls over financial reporting and report on its assessment. This mandatory report on internal controls has to be included in the company's annual report and requires explicit disclosure of the following:[25]

- Management's responsibility for establishing and maintaining adequate and effective internal controls over financial reporting.
- The framework used by management in its assessment of the effectiveness of the design and operation of internal controls over financial reporting.
- Management's assessment of the effectiveness of the design and operation of the company's internal controls over financial reporting.
- Disclosure of any identified material weaknesses in the company's internal controls over financial reporting.
- Disclosure that the company's independent auditor has issued an attestation report on management's assessment of the effectiveness of the internal controls over financial reporting.

As is clear from these disclosures, management bears much of the responsibility for designing and maintaining an effective internal control system. The responsibilities of the management are assigned by the government and enforced by the regulators. Unlike the other dimensions of corporate governance, which are either advisory (such as board of directors), subservient (such as internal audit and employees), or observer (such as external auditor), the management has the authority over operations and policies of the organization. Thus, the fair presentation of financial statements is ultimately the responsibility of the management. Accordingly, the management is responsible for the prevention and detection of fraud and misappropriation.

Management is the component of the corporate governance system that has the greatest motivation as well as opportunity to commit fraud. In that sense, management is often termed as the "Achilles heel" of the corporate governance system. Corporations' reward and compensation plans, which are typically linked to creating shareholder value or reported income, create a strong financial motivation for managers to engage in financial statement fraud. Often, management is under tremendous internal and external pressure to show more favorable performance and financial results. The management is under constant pressure of not only "meeting the numbers" but exceeding analysts' earnings expectations. Inability to show good corporate performance may be perceived as a personal failure for an otherwise highly successful individual. The pressure to not fail is sometimes a greater motivator than financial rewards in inducing the management to mis-state or obfuscate financial reporting.

According to the Center of Audit Quality, most major financial statement frauds historically involve senior management.[26] The management is in a unique position to perpetrate fraud by overriding controls and acting in collusion with other employees. Given that management exercises operational control over all aspects of the organization, it has the highest opportunity to commit fraud as well as conceal it. Management also has the opportunity to override controls that were put in place to prevent fraud. It creates a great challenge when the party responsible for design, operation, and evaluation of the controls chooses to override those. All other groups involved with corporate governance rely on the management for information and trusts it to follow the policies it designed and enforces. Hence, these other parties have to exercise professional skepticism and be cautious that management may in fact be the one committing fraud.

Management is also responsible for choosing accounting policies as well as making estimates inherent in financial reporting. It is expected that management will choose accounting policies that best reflect the underlying economics of the transaction and make good faith estimates. Sometimes, however, management may be motivated by factors other than faithful representation in their choices of accounting policy or estimates. Often times, management may engage in a practice known as *earnings management*, thus willfully choosing accounting policies and estimates to report earnings that fit their objectives. This is a much debated practice in academic research, and no consensus exists. However, when the "gamesmanship" of earnings management goes a bit too far, it can result in fraudulent financial reporting. A thin line separates legally acceptable but ethically dubious earnings management from illegal fraudulent financial reporting.

4.7 The Role of the Internal Auditor

The internal audit function is considered an important part of an effective corporate governance structure. The Institute of Internal Auditors highlighted the role of internal auditors in promoting an ethical culture in an organization and its active role in detecting misappropriation of an organization's assets.[27] It is evident that the objective of internal auditing is to support and strengthen an organization's corporate governance mechanism as well as evaluate the effectiveness of its risk management and control procedures. Internal auditors have made significant contributions to organizations through their systematic, objective, and disciplined approach to investigation. The internal auditors interact with all other parties of corporate governance in a meaningful way. This is the part of the structure that undertakes operational functions on a regular basis. The effectiveness of the system rests not only with its design and intent but also its implementation, and the internal auditors are the ones responsible for the latter.

SOX imposed restrictions on U.S. firms to outsource their internal audit function to the external audit firm. These restrictions were put in place to enhance auditor independence, both in reality as well as in perception. These restrictions imposed by SOX increased the organizational reliance on internal auditors. Specifically, internal auditors play an important role in compliance with Section 404 of SOX, and many adjustments have been made in response to it. Consequently, internal audit resources have been increased in many organizations in line with the increased demand and responsibility placed on internal audit services.

From a corporate governance standpoint, internal auditors are in the best position to prevent and detect financial statement frauds as well as asset misappropriations. Unlike external auditors, internal auditors are in place throughout the organization and not just at year-end and so can continuously monitor the internal control structure and assess failures on a timely basis. Being an employee of the organization, the internal auditor usually has a longer tenure than the external auditor, hence increased familiarity with personnel, management policies and procedures, and changes thereof. Furthermore, other employees of the organization may find internal auditors more easily approachable in sensitive and ambiguous situations rather than reaching out to outside parties or external auditors. Thus, internal auditors play a proactive role in organizations in preventing and detecting frauds. In fact, one of the largest frauds in the U.S., that of WorldCom, was unraveled by the internal audit team, details of which are presented in Exhibit 4.3.

Exhibit 4.3: Internal Auditors Unraveled Fraud at WorldCom

Cynthia Cooper, the then Vice President of internal audit at WorldCom, decided to investigate anomalies in the company's accounting entries that resulted in uncovering one of the largest fraud schemes the U.S. had ever seen. Her suspicion arose when the CFO asked her to delay the capital-expenditure audit and was initially dismissive of her findings. Even the controller admonished her for wasting time auditing capital expenditures. Despite the roadblocks, Ms. Cooper and her team continued to investigate and started working after hours so as not to be detected. Armed with evidence, they started approaching personnel in the accounting department, and finally the Controller confessed.

In this case, the reporting structure of the internal audit function presented a conflict of interest. Many chief audit executives still report to the CFO, who determines their compensation. For greater independence she suggests that internal audit function should directly report to the audit committee.

Professional guidance[28] for internal auditors defines the role of internal auditors in preventing fraud through the following three steps:

- Identification of red flags that signal the possibility of fraud.
- Investigation of the symptoms of fraud.
- Reporting of the findings to the audit committee and other appropriate levels of management.

The guidance also underscores the importance of internal auditors maintaining independence, objectivity, and authority in order to function effectively. The specific recommendations for internal auditors are

- Be respected within an organization.
- Have unrestricted access to records and information throughout the organization.
- Be authorized to communicate directly with the audit committee or the BOD.
- Have reporting lines and adequate resource approved by the Board and not solely by top management.

The Institute of Internal Auditors standard 1210.A2 requires internal auditors to possess "sufficient knowledge" to identify indicators of fraud. Their responsibilities for detecting fraud are also outlined in the Standard as

- Obtaining of sufficient knowledge and understanding of fraud risks to be able to identify "red flags" that fraud might have occurred.
- Obtain understanding of the corporate structure to identify opportunities for fraud due to lack of or weakness in internal controls.
- Ability to evaluate actions taken by perpetrators to commit fraud as well as conceal evidence.
- Access to appropriate authorities within the organization to inform of the possibility of the occurrence of fraud.

The internal auditor group is the one most likely and able to detect asset misappropriation fraud committed by lower-level employees. Because asset misappropriation committed by lower-level employees is likely to be immaterial in dollar amount to affect fairness of the financial statements, external auditors are unlikely to detect those. The internal auditors, on the other hand, are part of the organization, and it is part of their responsibility to ensure safeguarding of the organization's assets. They are sometimes expected to conduct standard forensic and fraud audit procedures to uncover employee fraud. Similarly, irregularities due to corruption or kickback schemes are more likely to be detected by internal auditors. Irregularities caused by corruption are more serious given they might violate laws in some jurisdictions. For example in the U.S. and U.K., the laws are stringent on bribery of foreign officials, and a violation of the law by one or few employees can make the entire organization liable. Thus, in detection of asset misappropriation and corruption, the internal audit function usually takes the lead role.

4.8 Who Blows the Whistle

As discussed throughout this chapter, fraud prevention and detection is a collective effort. It clearly is not the sole responsibility of just one party, but the responsibility is shared by many. However, it is still important to know how much of a role each of these parties plays in detection of fraud.[29] Interestingly, other constituents, employees, financial analysts, nonfinancial regulators and press, although not assigned traditional roles in corporate governance, have been responsible for providing protection against reprisals. The Dodd-Frank Act of 2010 in the U.S. attempts to limit reprisals against employees by enacting provisions that protect the whistle blower.

A recent academic study[30] gathered a comprehensive sample of 216 alleged corporate frauds in large U.S. companies between 1996 and 2004. For each of the instances the researchers discovered the party that first revealed the fraud, or the whistle-blower. According to the authors, fraud detection in the U.S. "relies on a village of whistle-blowers." Although internal governance or monitoring by the Board of Directors is highly effective and uncovered 34% of cases, the remaining were vastly dispersed in terms of which party detected it. Table 4.3 is adapted from the study and tabulates the raw data. Surprisingly, parties not typically considered players in corporate governance area played a key role in fraud detection. Employees uncovered 12% of the cases, financial analysts about 11%, nonfinancial regulators another 9%, and an unlikely group, the media, about 10%. These players are not typically considered key in discovery of fraud and are neither given responsibility nor the authority to do so.

Table 4.3 Who Detected Corporate Fraud (1996–2004)

Group Detecting Fraud	Number	Percentage
Internal Governance (BOD)	74	34.3%
Employee	26	12%
Analyst	24	11.1%
Media	22	10.2%
Industry Regulator	20	9.3%
External Auditor	16	7.4%
SEC	10	4.6%
Other	24	11.1%
Total	216	100%

Data Adapted from: A. Dyck, A. Morse and L. Zingales, "Who Blows the Whistle on Corporate Fraud?" 2010 *Journal of Finance*, 65.

These findings have serious ramifications on corporate governance mechanisms and society. First, the researchers documented significant costs of whistle-blowing for employees. Second, they documented that the whistle-blowing party does not get significant economic benefit from their actions of revealing corporate fraud. Specifically, auditors, analysts, and journalists do not seem to gain much, and the employees seem to lose outright from whistle-blowing.

Exhibit 4.4: Insider Whistle-blowing

Matthew Lee at Lehman Brothers

In May 2008, Matthew Lee, a Senior Vice President in Lehman's Finance Division, submitted a letter to senior Lehman management in which he alleged various financial reporting practices that potentially violated Lehman's own code of ethics. Consequently, Lehman's audit committee identified and treated Lee as a whistle-blower. The audit committee instructed Lehman's internal audit group and external auditors to investigate Lee's concerns.

On June 12, E&Y's engagement partner and another member of the audit team interviewed Lee. The auditor's notes of the interview indicate that Lee verbally informed E&Y that Lehman had used Repo 105 transactions to remove $50 billion of inventory from its balance sheet for the quarter just ended and returned the inventory to its balance sheet approximately one week later.[31] On the day following this interview, E&Y met with Lehman's audit committee but did not inform the committee of Lee's allegations regarding the Repo 105 activity.

In June 2008, Mr. Lee was let go from Lehman as part of a wave of firm wide layoffs. Mr. Lee had been unable to find work as of 2012 since being laid off and ironically believes it was due to his association with the auditing department at Lehman. Lehman filed for bankruptcy in September of 2008.

James Bingham at Xerox

Mr. Bingham, an Assistant Treasurer at Xerox with 15 years of service, alleged that Xerox may have improperly booked $140 million in revenue and $80 million in profits in 1999. He informed the chief financial officer and two other senior executives on August 28, 2000, at the corporate headquarters in Stamford, CT. He told his superiors that there was a "high likelihood" that Xerox in recent years has issued "misleading financial statements and public disclosures." Mr. Bingham specifically criticized the practice of booking for services and supplies as well as "rosy" assumptions about the future value of leases in developing countries. A few days after his presentation, Mr. Bingham was fired from Xerox. The company cited his "disruptive and insubordinate behavior." Xerox however took the allegations seriously and presented those to the audit committee as well as to the company's external auditors for further investigation.

Subsequently, the SEC launched a criminal investigation into accounting practices at Xerox that led to a restatement of results and a record $10 million civil fine. As a result of the review, Xerox made adjustments to its balance sheet, which reduced its shareholder's equity by $137 million. The company acknowledged that it had misapplied certain GAAP principles but insisted that it did not record any fictitious transactions.

> As for Mr. Bingham, he hasn't been employed since he was fired from Xerox. In 1999, Mr. Bingham's pay exceeded $350,000. Mr. Bingham's attorney sums up his situation as, "Jim had a great career but would never get a job in Corporate America again."

Employees who blow the whistle on corporate fraud bear a disproportionate amount of personal cost, primarily due to the lack of organizational safeguards protecting them against vicious and sometimes malicious attacks from powerful organizations whom they are accusing of wrongdoing. In the U.S., the Federal Civil False Claims Act (or the qui tam statute) provides some protection and monetary reward to compensate for the risk and misery that these individuals undertake. When the fraud involves a monetary outcome for the government, the individual who brought forth the allegation is entitled to 15% to 30% of the money received by the government. However, in Exhibit 4.4 we discuss two specific cases where the personal costs to certain employees who blew the whistle on corporate fraud were enormous.

Management fraud is more likely to be detected by employee tips than by any other method according to Association of Certified Fraud Examiners and also from a survey conducted by PricewaterhouseCoopers.[32] Further, an ACFE study reports that "approximately half of fraud tips came through a hotline when that mechanism was available" and further, "63% of the hotline reports involved fraud by a manager or executive." The importance for corporations to have confidential whistle-blowing mechanisms as well as adequate protection to whistle-blowers is underscored by this and other similar findings. In fact, the recommendation of maintaining a confidential hotline was made in the addendum to SAS 99, issued by the AICPA in 2002. Institute of Internal Auditors in 2010 in Emerging Trends in Fraud Risks identified a tool for confidential reporting as one of the key components of a fraud management program. The Sarbanes-Oxley Act makes the audit committee specifically responsible for establishing and overseeing a confidential reporting mechanism. SOX provided some degree of protection to the whistle-blower by maintaining anonymity and confidentiality of the complaint. The Dodd-Frank Act provides safeguards against retaliation to the whistle-blower and directs the SEC to reward whistle-blowers. For such programs to have credibility and be effective, complaints against senior management should be investigated promptly by the audit committee.

Global implications of the study are that jurisdictions should adopt *qui tam* provisions of sharing proceeds from successful litigation with the parties that brought the allegations to the forefront. Better safeguards have to be provided to employees who blow the whistle given that this is the group that most often bears a high personal cost. Many of the successful employee whistle-blowers have publicly stated that their personal costs were significantly high and if given a chance would not do it again. The high personal cost for employees in the U.S., with a relatively strong judicial system and infrastructure, implies that it possibly is much more burdensome in developing economies without similar safeguards. Safeguard of employee whistle-blowers appears to be a key measure that has to be legislatively adopted to further strengthen corporate governance measures. The Dodd-Frank Act in the U.S. purports to do so. The common adage of "if you see something, say something" developed in the post-9-11 world has to be broadened to the corporate realm to effectively deter and detect corporate fraud.

4.9 Chapter Summary

This chapter discussed the roles and responsibilities of various parties involved in the supply chain of financial reporting in preventing and detecting fraud. Though the public perception of the responsibility of detecting fraud is limited to external auditors, other parties bear responsibilities as well. Both the authoritative guidance and regulation, all over the world, assign responsibilities to all parties. Additionally, empirical evidence suggests that multiple constituents have uncovered fraud in the U.S. All these point to the need for shared responsibility in detecting fraud.

Each of the parties may work individually to deter and detect fraud; however, the collective sharing of ideas and information can lead to a more timely detection of fraud. To expedite collective sharing of ideas, it is imperative that there be formal lines of communication available to these parties. The communication lines in a typical organization are proposed in Figure 4.2. The solid arrows indicate the normal channels of communication, and the dashed arrows indicate additional channels for efficient and effective sharing of critical information. Management typically bears responsibility for cultivating an open communication culture in which employees feel empowered to directly communicate across all levels of the organization. Similarly, the Board of Directors and its Audit Committee might seek to interact with not only managers but other employees as well as major customers and vendors to enhance their knowledge about the company's operations and to be better able to assess potential risks. External auditors, for their part, could proactively seek opportunities to interact with the audit committee and senior management to discuss their audit plan and findings.

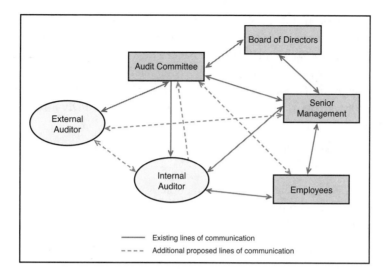

Figure 4.2 Lines of communication for effective corporate governance

Fostering a collaborative and information sharing environment could potentially not only lead to early discovery of fraud but may also act as a deterrent. One of the potent criticisms of the Examiner of Lehman Bankruptcy, Mr. Valukas, was the "silo" mentality that seems to have existed within the corporate governance structure of Lehman. Different parties implicitly relied on each other's professional judgment without explicitly discussing their misgivings or skepticism. Fostering communication and collaboration across different parties involved in corporate governance is the first step to abolishing the silos that may exist within the corporate governance structure. In summary, all associated individuals suffer monetary, reputational, psychological, and other damages when a fraudulent scheme implodes, hence it is imperative that all work collaboratively to prevent and deter such unfortunate occurrences.

4.10 Endnotes

1. Dale Flesher, and Tonya Flesher, Ivar Kreugers' Contribution to U.S. Financial Reporting, 61 *The Accounting Review*, 1986.

2. More recently on page 3 of the Proposal for a Regulation of the European Parliament and of the Council on specific requirements regarding statutory audit of public interest entities, available at http://ec.europa.eu/internal_market/auditing/docs/reform/regulation_en.pdf.

3. In the U.S., these include the Cohen Commission (1978), Metcalf Committee (1976) and Treadway Commission (1987). In the U.K. these include the Cross Committee (1977), Greenside Committee (1978), Financial Reporting Council (2005). In Canada these include the Adam Committee (1977) and MacDonald Commission (1988). More recently, the Green Paper issued by the European Union, http://eur-lex.europa.eu/LexUriServ/LexUriServ.do?uri=COM:2010:0561:FIN:EN:PDF, also addresses the expectations gap and the U.S. study.

4. Sweeney, B. (1997) "Bridging the expectation on shaky foundations," *Accountancy Ireland*, pp. 18–19.

5. D. Lindsay, "Auditor Client conflict Resolution: An Investigation of the Perceptions of Financial Community in Australia and Canada," from *International Journal of Accounting*.

6. R. Chowdhury, J. Innes and R. Kouhy, "The Public Sector Audit Expectation Gap in Bangladesh," from *Managerial Auditing Journal*.

7. ibid.

8. R. Dixon, A.D. Woodhead and M. Soliman, , "An Investigation of the Expectation Gap in Egypt," from *Management Auditing Journal*.

9. Mahdi Salehi, Ali Mansoury and Zhila Azary, "Audit Independence and Expectation Gap: Empirical Evidences from Iran," from *International Journal of Economics and Finance*.

10. M.N. Fadzly and Z. Ahmad, "Audit Expectation Gap: The Case of Malaysia," from *Managerial Auditing Journal*.

11. S.B. Adeyemi, and O.M. Uadiale, "An Empirical Investigation of the Audit Expectation Gap in Nigeria," from *African Journal of Business Management*.

12. P.J. Best, S. Buckby and C. Tan, "Evidence of Audit Expectation Gap in Singapore," from *Managerial Auditing Journal*.

13. T. Swift, and N. Dando (2002) "From methods to ideologies: closing the assurance expectations gap in social and ethical accounting, auditing and reporting," from *Journal of Corporate Citizenship*.

14. M. Salehi, and V. Rostami (2009) "Audit expectation gap: International evidence," from *International Journal of Academic Research*.

15. B. Porter, "An Empirical Study of the Audit Expectation-Performance Gap," from *Accounting and Business Research*.

16. Entire transcript of the discussion is available at http://c0403731.**cd**n.cloudfiles.rackspacecloud.com/collection/programs/sechistorical-102610-transcript.pdf.

17. Auditing and Assurance Standards Board, 2006.

18. The Advisory Committee on the Auditing Profession issued its final report on October 6, 2008 which makes recommendations on auditor responsibility to detect fraud. The entire report is available at http://www.treasury.gov/about/organizational-structure/offices/Documents/final-report.pdf

19. Page 7 of the Proposal for a Regulation of the European Parliament and of the Council on specific requirements regarding statutory audit of public interest entities available at http://ec.europa.eu/internal_market/auditing/docs/reform/regulation_en.pdf.

20. The CFA survey available at http://www.cfainstitute.org/memresources/monthlyquestion/2008/february.html.

21. The speech in its entirety is available in the archives at www.sec.gov date February 8, 1999.

22. Information summarized from "Audit Committee Guidance for European Companies" a KPMG publication dated September 2011 and sponsored by the European Confederation of Directors' Associations. Available at http://www.ecoda.org/docs/Publications/ecoDa%20guidance%20FINAL.pdf.

23. Information obtained from Japan's Corporate Governance Principle dated October 2001. Document available at http://www.ecgi.org/codes/documents/revised_corporate_governance_principles.pdf.

24. Toby Bishop, Director of Deloitte's Forensic Center, suggested these guidelines in 2009 under the title, "Ten Tips for Preventing Corporate Fraud" available at www.vqginc.com/fraud.html. The list provided paraphrases the content of that document.

25. List compiled from the U.S. Securities and Exchange Commission's Interpretive Guidance on ICFR, issued in June 2007.

26. "Center and Audit Quality, Deterring and Detecting Financial Reporting Fraud."

27. Institute of Internal Auditors, Practice Advisory 2130-1: *Role of the Internal Audit Activity and Internal Auditor in the Ethical Culture of an Organization*, 2004.

28. This paragraph is based on the Statement of Internal Auditing Standard No. 3: *Deterrence, Detection, Investigation and Reporting of Fraud*.

29. KPMG conducts a biannual fraud survey. The survey results, available on its website, are quite insightful and show interesting variation across geographic regions.

30. A. Dyck, A. Morse and L. Zingales, "Who Blows the Whistle on Corporate Fraud," from *Journal of Finance*.

31. Valukas 2010, Volume 3, p. 956

32. The survey is titled *Economic Crime in a Downturn*, which estimated 48% of fraud discovered was as a result to tips or calls to the hotline.

5
Data Mining

We are drowning in an ocean of data while starving for knowledge.
—John Naisbitt (*Megatrends*, 1982)

5.1 Introduction

Data mining or knowledge discovery in databases (KDD) is a collection of techniques and algorithms for unraveling patterns concealed in large data-sets. There are voluminous amounts of data in accounting, hence having scalable and efficient methods for finding patterns in data and relationships between data elements is of high value. The abundance of data in accounting, coupled with the need for powerful data analysis, is often described as a *data rich but information poor* situation. As the amount of available data increases, manual interpretation of data or assessing information from data has become more tedious, hence the advent of automated data mining tools. The knowledge discovery process from the data is illustrated in Figure 5.1. First, data from the database is selected and transformed if necessary. Next it is partitioned into two sets, the training set and the holdout set. The data from the training set is mined using various tools and techniques discussed in this chapter. The resultant model is tested for accuracy using the holdout set. The model is tweaked and the inaccuracies addressed.

Data mining tasks can be classified into two categories based on the purported use of the pattern. These two categories are descriptive and predictive. *Descriptive tasks* characterize the general characteristics of the data and, as the name suggests, describe the data with the aid of charts, graphs, and other visual aids. Predictive tasks, on the other hand, help draw inferences from data to identify patterns and trends.

In a forensic accounting context, the accountant may not have a priori knowledge of interesting patterns in the data, instead may discover those through repeated searches.

The process of data analysis in the context of forensic accounting is illustrated in Figure 5.2. Accountants usually encounter multiple databases and legacy systems in which a client's accounting data is stored. The first task of accountants is to clean the data, then to eliminate redundancies and data incompatibility. Next comes selecting task-relevant data. With increases in the volume of data to be analyzed, the data mining tools discussed in this chapter could provide interesting patterns and information to the accountant that aids in their investigation. Of course generating knowledge from patterns and subsequently presenting the knowledge as evidence relies upon the expertise of the forensic accountant. This chapter presents data mining tools to create patterns from data; later chapters discuss probability concepts that transform knowledge into evidence.

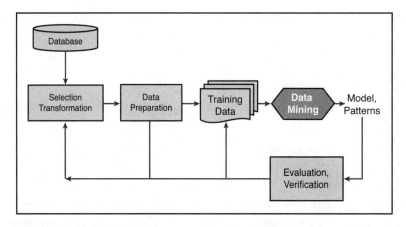

Figure 5.1 Steps in the Knowledge Discovery in Databases (KDD) process

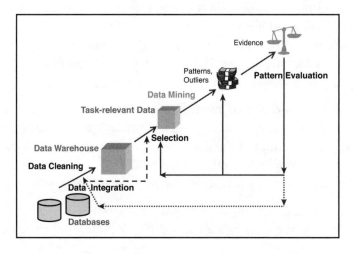

Figure 5.2 Data evolution process in forensic accounting

The forensic accountant usually has no check list or standard procedures to perform; instead, they are looking for interesting and unusual data patterns. The accountant looks for different kinds of patterns in parallel. Thus, the forensic accountant needs access to powerful data mining tools that can mine multiple kinds of patterns simultaneously. Additionally, data mining systems used by forensic accountants should be able to discover patterns by analyzing the data, patterns that the forensic accountant may be unaware of. At the same time, these systems have to be flexible to allow the forensic accountant to guide the search for patterns.

Data mining, as a discipline, is fast growing and rapidly innovating. The goal of this chapter is not to provide a comprehensive coverage of the extant techniques or the underlying methodology. Instead the goal is to familiarize you with this growing field and introduce the widely used techniques and concepts in data mining. The focus of the chapter is to illustrate that the powerful data mining tools for discovering patterns in data can be employed in a forensic accounting context to aid in probabilistic or statistical inference by relating common concepts in data mining to corresponding probability concepts.

The tools of data mining are powerful and can potentially generate many patterns, but the expertise and experience of a forensic accountant is invaluable in identifying which of these millions of patterns need to be further investigated. Typically, not all patterns detected by data mining tools are interesting or informative. An interesting pattern represents knowledge or supports a hypothesis.

First in this chapter, you learn about the elements of data classification for nominal variables and that of prediction for continuous variables. A discussion of the data mining tool, association analysis, in which patterns of variables that occur together are identified, follows. Next the tool and methodology of cluster analysis is presented, in which data points that are similar to others are clustered in a group to increase intra-group similarities while reducing inter-group similarities. Cluster analysis enables one to identify outliers. The chapter wraps up with an extended discussion of other methods and tools that can be used to identify outliers in the data.

5.2 Data Classification

Classification is the process of identifying a set of models that describe or distinguish between data elements or concepts. The derived model helps elicit simple "if-then" rules or portray the relationships using a decision tree or neural network formalism. Classification is a form of data analysis that is used to extract models from data that categorizes or classifies variables. It could also be used to determine the values of continuous variables. When used in the latter application, it is termed *prediction*. That is, when a task involves placing variables in finite values, it is termed classification, but when the task involves continuous variables or those with infinite values, it is called prediction. For example, determining whether an individual is credit-worthy is a classification problem, but determining the credit score is a prediction problem.

Data classification employs many probability-based techniques such as decision tree induction, Bayesian belief networks, neural networks, case-based analysis, rough sets, genetic algorithms, and fuzzy logic among others. An extensive discussion of these methodologies is beyond the scope of this book, but a summary description of how such techniques could be relevant for forensic accountants is provided. Traditional statistical techniques, such as linear regression models, are also applicable, which are discussed in a later chapter.

Data classification is usually a two-step process. The first step involves generating the model and specifying relationships between variables, also known as the training phase. The second phase involves validating the model and fine tuning it to improve accuracy, known as the testing phase. Prior to undertaking a classification task, the data is partitioned into two sets, the training set and the holdout set. The model is specified by using the data in the training set and its accuracy validated by testing it on the holdout set.

The training phase of the model could be either supervised or unsupervised. In a supervised setting the classes are prespecified, whereas in an unsupervised setting the class labels or even the number of classes is not specified or known in advance. Typically the trained model is represented in the form of classification rules, decision trees, or mathematical formulas. The result is then validated by determining its predictive accuracy on the holdout sample.

The classification analysis is indirectly useful in the context of forensic accounting. This analysis enables the auditor to generate expectations of a normal pattern of transactions in a business. When the expectations are formed and validated, significant departures from these expectations have to be investigated to ensure that no impropriety has occurred. This methodology is especially pertinent when large amounts of data are generated in a short span of time, which makes it almost impossible for an auditor to validate each and every transaction. The classification analysis can assist the auditor in directing the auditor's attention to those cases, which does not comply with the self-generated rules of the classification system.

5.3 Association Analysis

Association analysis is the discovery of association rules that occur frequently in a given data-set. Association analysis is especially relevant for transaction data analysis where various attributes are linked. Association rule mining searches for interesting relationships among items in a given set. The earliest form of association rule mining was in the area of "shopping cart" analysis, which is an investigation of buying patterns of consumers.

The *shopping cart* or *market basket* analysis is one of the earliest uses of data mining techniques. The buying habits of consumers are empirically established, and the results are used to plan marketing and advertising strategies. For example in a grocery store, items that are regularly purchased together are placed in close proximity in order to further encourage joint sales of such items. An illustration of a market basket is shown in Figure 5.3. Although customer purchases at a grocery store vary, relevant patterns can be found. For example, not many consumers purchase baby food, but those who do have a higher propensity of buying diapers as well. Hence, baby food and diapers are placed in close proximity at stores. Similar purchasing patterns can be identified, and the complementary items are placed near each other.

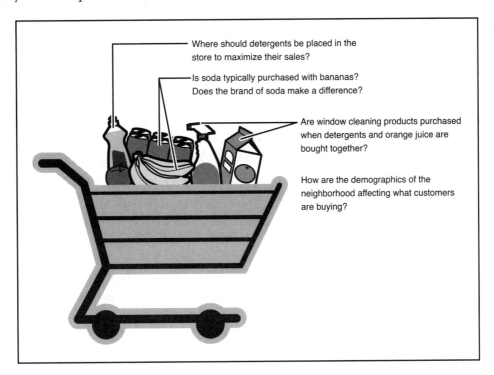

Figure 5.3   Association analysis of market baskets

The data mining technique associated with association analysis is based on Boolean vectors. In Boolean algebra each variable can take only one of two values, 0 or 1. In the case of a retailer, each item can be denoted by a Boolean variable, and each consumer purchase can be treated as a vector of 1s and 0s, with 1 being assigned to the Boolean variable corresponding to the item purchased and 0 otherwise. As each consumer order or purchase leads to a Boolean vector, most retailers will collect data on billions of Boolean vectors across all of their locations. These Boolean vectors can then be analyzed to find association between different items. In other words, the Boolean vectors help identify complementary and substitute items. Complementary items are those that are usually purchased concurrently, and substitute items are never purchased concurrently.

The patterns identified here are represented as association rules. Rule support and rule confidence are two metrics that measure the attributes of the rule. *Rule support* indicates the usefulness of the rule, and *rule confidence* reflects the certainty of the discovered rule. These metrics are reported as percentages or as measures between 0 and 1. A rule support of 0.04 indicates that the joint purchase was undertaken in 0.04 of all the purchases. A rule confidence of 0.75 indicates that given a consumer purchased the first product, there is a probability of 0.75 that they also purchased the second product. In the next chapter, these numbers are defined in terms of joint and conditional probabilities. Rule support will be defined as joint probability, and rule confidence will be defined as conditional probability. These probability measures, essential to using statistics, can be obtained in a real-world setting through data mining, provided there is a large amount of data to process as in the marketing example.

Association analysis can be used in a forensic accounting context to unravel collusion between employees. Properly designed internal control systems in an organization prevent theft of assets by an employee. However, if two or more employees collude, they could subvert a properly designed internal control system, and their theft or asset misappropriation will remain undetected.

Consider for example employees stealing customer payments made to a company. Most organizations institute proper internal controls over their cash receipt process by assigning different employees to deposit checks, make journal entries, or write off customer accounts. For a large organization, there may be multiple individuals assigned to each one of these roles. Theft of customer payment by any of these employees will soon be detected through the in-built system of checks and balances. The properly designed system, however, could be rendered ineffective by employee collusion. That is, for theft to take place, all three employees must be working in the same shift. An association analysis of employee work shifts would unravel any unusual association.

The association analysis to detect collusion across segregated employee responsibility can be designed as follows. In the scenario just given, of 3 employees colluding to steal customer payments and fraudulently writing off the receivable, suppose there are 5 employees assigned to each of the 3 roles. That is, there are 15 employees in total. Further suppose the database for write-offs records the date, time, and amount of write-offs in addition to the employees signing off on the write-off. The 15 employees associated with partially authorizing a write-off can be represented by a Boolean vector of 15 elements, each element corresponding to an employee. For each write-off the Boolean vector is populated with a value of one assigned to the employees who authorized or "signed-off" on the write-off and zero otherwise. For any given write-off, 3 distinct signatures or authorizations are required, 1 employee from each of the 3 functions. Thus there are 5^3 or 125 unique vectors possible. The forensic accountant can compile the frequency of each of these 125 unique vectors in the population of write-offs using data mining techniques. Suppose in a year there were 1,000 write-offs of customer accounts. If one of the Boolean vectors, or a set of 3 employees, accounted for 400 of those write-offs, that would be perceived as a warning sign of collusion between those three employees. Probabilistically, because there are 125 possible combinations of employees out of 1,000 write-offs, you can expect each combination of employees to be responsible for 8 of those. Forty percent of the write-offs being authorized by the same set of 3 individuals should raise a red flag. Forensic accountants can then investigate such unusual associations to ascertain whether there was any collusion between the employees.

The technique of association analysis could be used to estimate many statistical measures that are discussed in subsequent chapters. Association rule mining consists of association rules of the form "if A then B." These rules are then evaluated based on the level of support and the level of confidence. Statistical tools such as correlation analysis could be used to provide statistical validity to the rules. These rules, when statistically valid, provide important conditional and joint probability matrices, which are helpful in the evaluation of evidence, as discussed in the next chapter.

5.4 Cluster Analysis

Cluster analysis is performed on a group of similar data elements to distinguish between dissimilar data elements. The objective of cluster analysis is to increase homogeneity within a cluster and increase heterogeneity across clusters. The clusters of objects are formed so that the objects within a cluster have high similarity to one another but are highly dissimilar to objects in other clusters. Formally stated, the objective of cluster analysis is to *maximize intraclass similarity while minimizing interclass similarities*. Each cluster that is formed can be viewed as a class of objects from which rules can be derived. Cluster analysis has its roots in multiple disciplines including statistics, biology, and machine learning.

The process of grouping a set of physical or abstract objects into classes of similar objects is a natural human trait. Even in early childhood, infants learn to differentiate between plants and animals, cats and dogs, edible items and non-edible items, through an iterative process of clustering. As infants grow older, their clustering becomes more and more refined, and they make fewer mistakes in recognizing patterns. Additionally, biologists and evolutionists have long used clustering to develop taxonomies and evolution processes of plants and animals. Hence, the idea and methodology of clustering was well studied in biology and human development much earlier than its use in business applications.

Clustering enables identification of dense and sparse regions in a data-set and therefore facilitates the discovery of the overall distribution patterns and correlations among data attributes. Clustering is an example of unsupervised learning in that the classes or labels are not pre-specified instead these are determined by the data. In that sense, clustering can be thought of as learning by observation or learning through experience. Data mining efforts have been concentrated on finding efficient and effective methods to identify clusters in large databases.

One of the common methods of clustering from a random set of data is illustrated in Figure 5.4. Panel A of the figure shows a box plot of the data points that seem randomly distributed. In the process of clustering, the first step is to specify how many clusters are desired, say three for our example. Next, three data points are randomly picked, which are marked with a * in Panel B and are called centers. Other data points are distributed to the three clusters based on their proximity to the centers. Such a distribution forms envelopes shown through a dashed line in Panel C. Now the mean value of each cluster is recomputed, and that forms the new *cluster center* in Panel D, and all other data points are reassigned to the three clusters based on their proximities to the new center. This process iterates until no further redistribution occurs and the process terminates with identification of final clusters shown in Panel E of Figure 5.4.

When the number of clusters is not prespecified as in the previous example, a different algorithm to determine the clusters is used, which is illustrated in Figure 5.5. This method is called the Chameleon.[1] The initial stage of Figure 5.5 shows the data points in a box plot. In the intermediate phase, each point is connected to its nearest neighbors, hence the plot depicts a nearest-neighbor graph. The nearest neighbor graph captures the concept of neighborhood dynamically, and the neighborhood radius is determined by the density of the region in which the object resides. The radius is small if the object resides in a dense area but is large if the object resides in a sparse region. That is, the neighborhoods are defined narrowly in dense areas but are defined widely in sparse areas. This process tends to evolve into more natural clusters. The relative interconnectivity and relative closeness are two metrics that determine similarity between each pair of clusters. It has been empirically established that such an approach is better suited at discovering arbitrarily shaped clusters.[2]

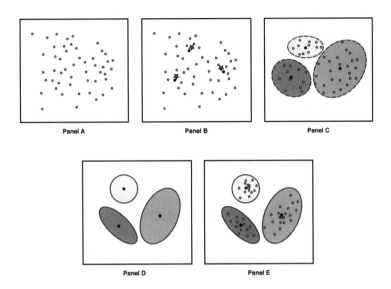

Figure 5.4   Illustration of clustering in a set of data points

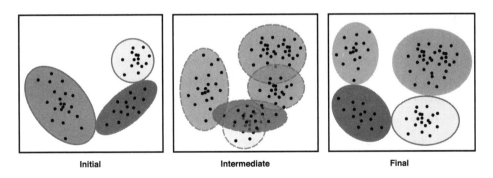

Figure 5.5   Illustration of the process of clustering

Chapter 5 Data Mining   97

5.5 Outlier Analysis

In some situations a database might include data that does not conform to the general rule derived for the dataset or the general behavior of other data elements. Such data elements are termed *outliers*, a term commonly used in probability and statistics. Although most common applications of data analysis tend to disregard and ignore the outliers, these have a special significance in forensic accounting. In applications such as fraud detection, these rare events or outliers may be more interesting and informative than the more regularly conforming events. The analysis of outlier data, relevant for fraud detection, is referred to as outlier mining.

Outliers may be detected using statistical tests that assume a probability distribution for the data and a distance measure such as standard deviation that identify elements that are significantly apart from the cluster. Alternatively, instead of using probability models and statistical concepts of distance, outliers can be identified by examining the differences in the main characteristics of objects in a group. For example, outlier analysis is often used to identify fraudulent use of credit cards. Say nine transactions on a particular credit card were undertaken in New York City and a cab fare in Chicago; the cab fare in Chicago does not conform to the general characteristics of the rest of the data and is identified as an outlier, requiring further investigation. Examples of statistical methods to identify outliers would be to compare the number of transactions of a given period. Mean and standard deviation of the number can be estimated from historical data, and an excessive number of transactions over a given period can be quickly identified.

The formal approach to outlier mining is a two-step process. In the first step normal patterns or relationships are established using the data-set. In the second step, significant departures from the normal pattern or relationship are identified. In a regression model, as is discussed in a later chapter, analysis of residuals provides a good estimation for data abnormality. For two or three -dimensional data-sets, plotting the data and visually identifying the outliers could be an effective and efficient approach. However, visual determination is not possible in situations when there are four or more attributes to each data point. Further, for non-numeric or categorical data, defining outliers could be problematic.

Figure 5.6 illustrates identification of outliers through applying cluster analysis techniques. The cluster analysis algorithm is modified slightly by putting in a constraint of maximum distance of each point in the cluster from the center. When such a measure is specified, observations that fall far from the center of respective clusters are isolated and identified as outliers.

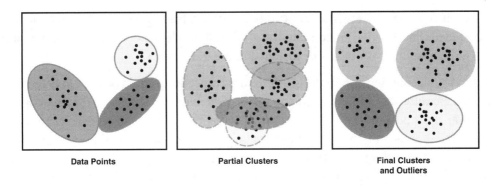

Figure 5.6 Identification of outlier through clustering

Methods of outlier detection could be parametric or nonparametric. *Parametric* methods use tools that have a basis in statistical theory and make an underlying assumption regarding probabilities. These methods flag as outliers those observations that deviate from model assumptions in a "statistically significant" way. Parametric methods, however, have limited application in new data-sets where there is a lack of prior knowledge of the parameters or distributions critical to applying such methods. Most of the parametric methods require the distribution of data to be identical and independently distributed. Needless to say, these assumptions are often violated in real world data. In those instances, nonparametric methods, which do not make probabilistic assumptions, could be used. *Nonparametric* methods are usually distance-based methods in that they measure distance from each observation to its nearest neighbor. When the distance between an observation and its nearest neighbor exceeds a pre-determined amount, the observation is deemed an outlier. Use of clustering approaches could lead to detection of outliers as discussed in the previous paragraph and illustrated in Figure 5.6.

Outlier detection methods can use a single-step or a sequential approach. In a single-step procedure all outliers are identified at once. In a sequential approach, at each step one observation is tested for being an outlier. The process continues until no remaining observation can be regarded as an outlier. Sequential approach can be further classified based on inward testing and outward testing. In inward testing, or forward selection methods, at each step the most extreme observation, the one with the largest metric specified, is tested for being an outlier. In outward testing procedures, the sample of observations is first reduced to a smaller sample while the remaining observations are kept in a reservoir. Statistics are calculated for the reduced sample. These statistics are then used to test the data in the reservoir to indicate whether these are outliers. The outliers are identified and removed from the data-set. Observations in the holdout sample not classified as outliers are incorporated back into the training set, and the statistics are recomputed. The iterative process continues until all the elements in the holdout set have been either classified as outliers or included in the training set.

For forensic accounting applications, any of these procedures could be used to determine outliers. Because the outliers form the focus of further investigation, the objective of outlier analysis is to reduce the number of observations to a manageable level but to be careful as not to accept an extreme observation as part of the normal population. That is, the risk of accepting an outlier as normal is more consequential than identifying a normal observation as an outlier. Outlier analysis is particularly relevant in continuous audit settings as well as for cyber-security applications.

5.6 Data Mining to Detect Money Laundering

Money laundering is a process of disguising the illegitimate origin of "dirty" money and making it appear legitimate. Through money laundering criminals try to convert money obtained from illicit activities into "clean" funds using a legal medium such as legitimate retail or investment banks. The objective is to conceal the original source of funds, location, ownership, and control of the money. The U.S. Department of Justice defines money laundering as

> ...the process by which criminals conceal or disguise the proceeds of their crimes or convert those proceeds into goods and services. It allows criminals to infuse their illegal money into the stream of commerce, thus corrupting financial institutions and the money supply, thereby giving criminals unwarranted economic power.

As money laundering activities are prevalent in drug trafficking, arms dealing, and terrorism, these pose a serious threat not only to financial institutions with which they transact but also to various countries. In response to the serious impact of the crime, governments and financial regulators mandate that financial institutions implement processes and procedures to deter and detect money laundering schemes. For example, in 2011 the FBI in the U.S. investigated 303 cases resulting in 37 indictments and 45 convictions in money laundering cases.[3] Recent settlements for money laundering charges in the U.S. against large financial institutions are summarized in Exhibit 5.1.

> **Exhibit 5.1: Recent Settlements for Money Laundering Charges**
>
> In December 2012 HSBC, Europe's largest bank, agreed to pay $1.92 billion in settlement of money laundering charges against it.[4] This was the largest penalty ever imposed on a bank. The U.S. government accused HSBC of transferring funds through the U.S. for Mexican drug cartels and also for Iran, which is under international sanction. Specifically, HSBC was accused of transferring illegal drug proceeds of $7 billion in cash between U.S. and Mexico. Although the amount of transaction raised certain red flags within the bank, according to reports those were waived under the pretext that this was primarily money being sent home by Mexican landscapers working in the U.S.[5] HSBC's U.S. division was accused of providing money and banking services to banks in Saudi Arabia and Bangladesh thought to have helped fund al-Qaeda and other terrorist groups. Also in December 2012, Standard Chartered PLC, another British bank, settled a money laundering investigation with U.S. regulators by agreeing to pay $470 million. The charges against it were money laundering allegations involving Iran. The penalties for both banks were approximately 10% of their 2012 pre-tax profits.
>
> In the summer of 2012, a U.S. Senate investigation concluded that HSBC's lax controls exposed it to money laundering and terrorist financing.[6] In its settlement, HSBC conceded that its anti-money laundering measures were inadequate and that it has taken steps to "beef up" its internal controls. The bank would also continue to strengthen its compliance policies and procedures, and its progress will be evaluated by an outside monitor appointed by the Department of Justice. The bank was accused of violating the Bank Secrecy Act and the Trading With the Enemy Act.
>
> Investigation of money laundering by banks has become a priority for U.S. law enforcement agencies. Since 2009 many prominent banks have been charged and have settled allegations that they laundered money for people, nations, or companies that were on the U.S. sanctions list. The list of banks that have paid substantial settlement since 2009 includes Credit Suisse, Barclays, Lloyds, and ING.

Traditional approaches to anti–money laundering initiatives followed a labor intensive manual approach.[7] However, given the volume of banking data and transactions, manual approaches have to be supported by automated tools for detecting patterns that are created by money laundering schemes. Furthermore, to detect money laundering and other financial crimes, it is important to integrate information from multiple databases. Given the enormity of the task, volume of data, multitude of sources, and rapidly changing criminal schemes, automated techniques are necessary to effectively detect emerging patterns.

One of the major challenges in detecting money laundering schemes is the unavailability of required data at any single database. The distribution of data-sets first requires a collection process and integration into a data warehouse. Because these various data-sets could be maintained by different authorities and be under control of different jurisdictions, getting permission to aggregate the data and compile a data warehouse is not a trivial task. The data on financial transactions consist of four layers: transaction, account, institution, and multi-institution. The most basic level is transactions, but they do not provide much context because they do not constitute links to accounts or other data. In the second level, multiple transactions are associated with an account. Analyzing accounts provides some information on the nature of financial activities in these accounts. When a customer has multiple accounts in the same financial institution, a consolidation of these accounts may enable the institution to analyze transfers across accounts and detect suspicious activities. The top level integrates a customer account across institutions, however linking accounts across financial institutions and perhaps regulatory jurisdictions may be impossible to achieve. Hence, a financial institution operates at a level below optimum to detect money laundering. Encouraged co-operation between institutions to share account information while useful in unraveling money laundering schemes may unnecessarily infringe on the privacy rights of the general population.

The task of anti-money laundering tools is to detect unusual patterns such as large amounts of cash flow at certain periods, by certain groups of people, to certain regions, using shell corporations, and so on. Multiple data mining tools discussed in this chapter can provide useful information. Data visualization tools can transform the data into graphs and charts that can display transaction activities at certain times or by certain individuals. Data classification tools can be used to identify attributes that are risky. Data association tools can be used to identify links among different people and activities. Clustering tools can be used to group different cases based on magnitude, region, individuals, and so on. Finally, outlier analysis tools can detect unusual amounts of fund transfers and other activities. These tools may identify important relationships and patterns of activities and help forensic accountants focus on a limited number of cases rather than be overwhelmed by the volume of information and data.

Although the use of data mining tools could ease the task of processing and aggregating large amounts of data, sole reliance on these tools to detect money laundering problems is not prudent. Data mining tools are constructed to detect patterns that appear in large sets of data and identify outliers that do not follow that pattern. However, money laundering schemes are not self-revealing; instead, much care has been exercised to masquerade these transactions as legitimate and normal transactions. Moreover, money laundering transactions are rare relative to normal transactions that banks process each day. Smart criminals with extensive knowledge of the financial process, who operate the money laundering schemes, design the transactions within the confines of a normal distribution so as not to raise suspicion. Hence, simplistic use of mean and standard deviation rules may not detect money laundering transactions; rather, more sophisticated and quick learning data mining tools are required.

5.7 Chapter Summary

The task of the forensic accountant is to search for exceptions, oddities, patterns and suspicious transactions in a data-set. Unlike financial audits, there are no standard procedures or check lists that a forensic accountant can use. Instead a forensic accountant relies on their judgment, intuition, experience, and the developed "sixth sense" to identify unusual circumstances that require further investigation. Data mining techniques help ease the task of forensic accountants by systematically reducing the set of observations or transactions that the forensic accountant must consider. Data mining is the process of discovering previously unknown and actionable trends, patterns, and relationships in the data-sets.

Some of the data mining tools were reviewed in this chapter. The purpose here was not to delve into mathematical details of these techniques but to make the reader aware of the relevance of these techniques to the task of forensic accountants. There are software packages that can be purchased commercially and used to assist the forensic accountant. The most relevant data mining tools for accountants are those that engage in discovery, predictive modeling, and deviation analysis. Although these tools are not a panacea for forensic accountants, use of these can aid the accountant by limiting the search space leading to focused attention on problem areas.

As discussed in the context of the anti–money laundering initiative, data mining tools, while helpful, are not perfect. That is, while these tools can potentially narrow the set of suspicious accounts and transactions, they do not identify these perfectly. There will be legitimate transactions included in the "suspicious" set, and some illegitimate money laundering transactions will be deemed normal. The discipline is still evolving, and the tools are becoming more sophisticated. Given the enormity of the task, the forensic accountant has little choice but to learn and adopt these tools in their investigations.

5.8 Endnotes

1. It was first proposed by G. Karypis, E.H. Han, and V. Kumar in "CHAMELEON: A hierarchical clustering algorithm using dynamic modeling" in *Computer*.

2. Jiawei Han and Micheline Kamber, *Data Mining: Concepts and Techniques*. Data from the FBI's *Financial Crimes Report to the Public,* available at http://www.fbi.gov/stats-services/publications/financial-crimes-report-2010-2011.

3. The settlement was announced by Senator Carl Levin, the Chairman of U.S. Senate Committee on Homeland Security and Governmental Affairs, available at http://www.hsgac.senate.gov/subcommittees/investigations/media/levin-statement-on-hsbc-settlement.

4. Based on an NBC News report available at http://www.nbcnews.com/business/report-hsbc-allowed-money-laundering-likely-funded-terror-drugs-889170.

5. A summary of the 330-page Senate report is available on the U.S. Senate website. The summary of the report can be accessed at http://www.hsgac.senate.gov/subcommittees/investigations/media/hsbc-exposed-us-finacial-system-to-money-laundering-drug-terrorist-financing-risks.

6. For further details see R. C. Watkins, et.al., "Exploring data mining technologies as tool to investing money laundering" from the *Journal of Policing Practices and Research: An International Journal.*

7. J. Kingdon, "AI fights money laundering," *IEEE Transactions on Intelligent Systems.*

6
Transitioning to Evidence

When you have eliminated the impossible, whatever remains, however improbable, must be the truth.
—Arthur Conan Doyle (Sherlock Holmes)

6.1 Introduction

When examining patterns derived from data mining, the pressing question is what do these patterns indicate? That is, what does the data suggest in relation to the propositions? When presented with information or data, a forensic accountant has to assess whether the information supports, negates, confirms, or disconfirms an assertion or hypothesis of financial impropriety. The forensic accountant's assessment of information and data is not limited to reporting the features of the data, otherwise known as descriptive statistics, but to demonstrate how the data or information supports one assertion or hypothesis more than the alternatives. When the data or information is used to validate one hypothesis or invalidate another, information transitions into evidence.

The concepts of the nature and the strength of evidence are of critical importance to the forensic accountant. Transitioning from information to evidence requires analyzing possibilities and "what if" scenarios. It requires eliciting possible causes and assessing the relationship between those causes and evidence accumulated. Failure to list all causes or misjudging relationships could lead to untenable conclusions.

This chapter presents the qualitative foundation for understanding properties of evidence. After introducing basic terminology and definitions a pictorial representation of the relationship is presented. Next, the notions of confirming evidence, supporting evidence, and strength of evidence are discussed. The link between information and evidence is drawn, and it is shown that not all items of information qualify as evidence in a forensic accounting context. The strength of evidence in support of an assertion is measured not by how likely the evidence is if the assertion were true, but more importantly by how unlikely the evidence would be if the assertion were false. Finally, the relevance, applicability, and importance of Bayes' Rule to the disciplines of forensic accounting and auditing is discussed.

6.2 Probability Concepts and Terminology

Probability is a numerical representation of uncertainty. The uncertainty arises from not knowing what the outcome would be. A *sample space* is a set of all possible outcomes. An *event* is any subset of outcomes within the sample space. Impossibility is denoted by ∅ a null event, and is the absence of any outcome in the set. Probability of an event or outcome takes a number between zero and one, or alternatively could be represented as a percentage between 0% and 100%. The extreme values denote certainty, a probability of zero indicates that we are sure that the event will not occur, or the assertion is false. Similarly, a probability of one denotes that we are certain that the event will occur or the assertion is true. Complete ignorance, or lack of knowledge, is represented by assigning equal or uniform probabilities to all possible outcomes or assertions.

A sample space consists of many events, and the item of interest might be the joint occurrence of two events, say A and B. This is represented by A∩B, A intersection B, and is a set of outcomes that belong in both sets, A and B. The term *joint probability* is used to denote the probability of the intersection of A and B. It is possible that events A and B have no common outcomes, or A∩B is ∅. When A intersection B is a null set, it implies that A and B cannot jointly occur, or they are *mutually exclusive*. Mutual exclusivity is an important concept in probability and has bearing on forensic accounting. The concepts of sample space, events A and B, joint probability, and mutual exclusivity are pictorially explained in Figure 6.1.

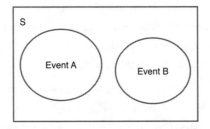

The rectangle S denotes sample space. The two events A and B are denoted by the two circles. As the circles do not intersect, the probability that A and B could jointly occur is zero.

Figure 6.1 An illustration of mutually exclusive events

When considering several events, it might be of interest to know that at least one of them will occur. This is denoted by the *union* of two events, denoted as AUB, and is the set of all outcomes that belong to at least one of the two events. Hence, AUB occurs when either A or B or both occur. The Venn diagram in Panel A of Figure 6.2 shows the union, which clarifies that an outcome is in AUB, if and only if it is in A or B or both. When the union of several events covers the entire sample space, these events are called *collectively exhaustive*. A set of basic outcomes that belongs in the sample space S but are not included in the event A is called the *complement* of A, denoted as ~A. This is pictorially represented in Panel B of Figure 6.2. By definition, an event and its complement are mutually exclusive and collectively exhaustive. For example, "heads" and "tails" are complements, in that these are mutually exclusive and collectively exhaustive events of a simple coin toss.

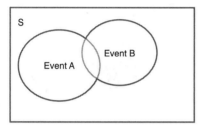

Panel A: Union of events A and B

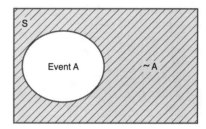

Panel B: Complement of event A, denoted as ~A

Figure 6.2 Venn diagram illustration of "union" and "complement"

This notion of sample space and events can involve two distinct sets of events, say A_1, $A_2, \ldots A_m$ and $B_1, B_2, \ldots B_n$. This formulation is applicable to decision-making in forensic accounting and auditing. The events A_i are mutually exclusive and collectively exhaustive within its set; similarly events B_j are mutually exclusive and collectively exhaustive within its set. The intersections between these events, $A_i \cap B_j$ are the (i × j) basic outcomes, which are called *bivariate*. In this context, the probability assigned to the joint event $(A_i \cap B_j)$ is the joint probability, and the probability for individual events is termed marginal probability. To compute marginal probability, we sum the corresponding mutually exclusive joint probabilities. That is,

$$P(A_i) = P(A_i \cap B_1) + P(A_i \cap B_2) + P(A_i \cap B_3) + \ldots + P(A_i \cap B_n)$$

similarly,

$$P(B_j) = P(A_1 \cap B_j) + P(A_2 \cap B_j) + P(A_3 \cap B_j) + \ldots + P(A_m \cap B_j)$$

6.3 Schematic Representation of Evidence

A diagrammatic representation of assertion and evidence in a sample space is possible through Venn diagrams, as illustrated in the previous section. Next is a continuing analysis of the relationships between assertion and evidence for additional scenarios shown in Figure 6.3. Panel A of Figure 6.3 depicts a scenario in which evidence E is confirming of Assertion A. Because E is a subset of A, if an outcome is in E, it must be in A, hence observation of evidence E in this case confirms the assertion A. However, Panel B depicts the situation in which the observation of evidence may fallaciously be presented as confirming the assertion. Evidence E is seen to occur even when the Assertion A is not true, thus observation of the evidence does not confirm the assertion; instead, ~E confirms ~A. That is, not finding the evidence E establishes that assertion A cannot be true.

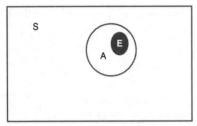
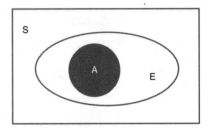

Panel A: E is confirming of A   Panel B: ~E is confirming of ~A

Figure 6.3 Illustrations of assertions and evidence

Confirming evidence is an item of evidence or a set of evidence, which when observed, proves that a proposition is true regardless of the initial belief on the assertion. The concept of confirming evidence has been formally studied in many disciplines including philosophy, mathematics, and law.

A confirming item of evidence is defined such that its observation confirms the assertion, regardless of how *small* the initial belief might have been. Panel A of Figure 6.3 shows a Venn diagram for a confirming item of evidence. In this case, if you observe the evidence, you can be sure that the assertion is true. For simplicity, suppose the assertion is "there is a cloud in the sky," and the evidence is, "it is raining." Because it can only rain when there are clouds in the sky, observation of rain confirms the presence of clouds. In this case, regardless of how small the prior belief was on clouds, evidence of rain could increase the probability of clouds from an infinitesimal amount to 100%. An item of evidence confirms an assertion if, and only if, the likelihood of obtaining the evidence given the negation of the assertion is zero. In this example, because probability of rain when there are no clouds is zero, observation of rain confirms the existence of clouds. It is important to note that the property that makes an item of evidence confirming is the impossibility of the evidence if the assertion isn't true.

In forensic accounting, a "smoking gun" is an item of evidence that by itself confirms an assertion. Rarely, however, in forensic accounting would any single item of evidence be confirming, or be a smoking gun. In general, a set of evidence, collectively, could confirm a proposition. The set might consist of items of evidence which, in isolation, are mildly supportive of the assertion, however in conjunction with other evidence, is confirming. This situation is shown via a Venn diagram in Figure 6.4.

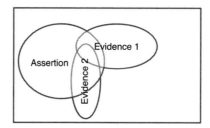

Figure 6.4   Schematic representation of confirming evidence

Consider the insider trading case by the Goldman Sachs Director that was discussed in Chapter 4, "Detection of Fraud: Shared Responsibility." Records of trading by the hedge fund showed an unusual trade of purchasing $43 million of Goldman Sachs stocks made at 3:56 p.m. on September 23, 2008. The stocks were sold the very next day, netting the hedge fund a profit of $1.2 million in a few hours of holding the shares. By itself the evidence of the trade, though unusual, is by no means confirming of any foul play. However, a related item of evidence was that a Director of Goldman Sachs had called the hedge fund manager three minutes prior to the trade, at 3:51 p.m., soon after the conclusion of a Board of Directors meeting at Goldman. Again the evidence of a Director making a call to a hedge fund manager right at the conclusion of a Board meeting is unusual, but by itself is not confirming of foul play. The two items of evidence together, however, present a scenario schematically shown in Figure 6.4. The jury in this case found the evidence sufficiently compelling to convict both the hedge fund manager and the Director of insider trading.

6.4 Information and Evidence

Knowledge or information does not necessarily constitute evidence. Evidence is related to an assertion, and observation or knowledge of it would make the assertion more or less likely. In situations when knowledge does not affect the belief or probability of the assertion, it is not considered evidence.

Consider the following example of a situation when information does not enable you to distinguish between two competing assertions. Suppose the assertion is it rained overnight, and we were informed in the morning that the grass is wet. Does the information translate to evidence regarding rain? Not necessarily, the grass could be wet if the sprinkler was on at night or because of dew. The information translates to evidence if it alters the prior beliefs on the assertions. For information to be considered evidence, it must be logically relevant. That is, it makes competing propositions more or less likely. Consequently, information is logically irrelevant if it does not change the likelihood of competing propositions. Whether or not the information alters the beliefs on competing assertions falls under the purview of the mathematical topics of probability and statistics.

In a forensic accounting context, when examining variations from normal patterns of data, you must consider the possibility that the deviation is due to chance and not reflective of a purposeful act. Evidence that is obtained has to be evaluated from the perspective of it being caused by chance or through normal variations of the business process. For example, higher than average recorded sales in the last week of December, by itself, is not evidence of financial misstatement. There is roughly a 50% chance that any particular week's sales is higher than average, and therefore by pure chance the last week of December sales may have been higher than average. A further detailed analysis is required to measure whether the deviation from average is within expected bounds or is it abnormal. Steps to conduct such analysis through statistical methodology are addressed in the remaining chapters. This chapter lays the foundation for statistical analysis by explaining common probability terminology and concepts.

6.5 Mathematical Definitions of Prior, Conditional, and Posterior Probability

Probability is quantification of uncertainty; it is the numeric value representing the chance, likelihood, or possibility of an event occurring or an assertion being true. Probability has a range between 0 and 1, both inclusive. An event that has no chance of occurring or an assertion that is deemed to be false is assigned a value of 0. Similarly, an event that will occur for sure or an assertion that is known to be the truth is assigned a probability of 1.

A-priori probability is the probability of occurrence of an event or the validity of an assertion based only on background knowledge of the process involved. The a-priori probability could be objective when based on empirical data or could be based on subjective assessment of the decision maker. The objective probability is derived through mathematical properties, such as probabilities in rolling a dice or through data gathering, such as surveys. Subjective probabilities, as the term suggests, differs from person to person. It measures one's belief on the validity of an assertion or occurrence of an event and is based on the individual's past experience, personal opinion, among other factors.

A simple event is described by a single characteristic, such as heads on a coin toss or a roll of 4 on the die. The complement of an event is every possibility except the event; that is, the complement of getting 4 on a roll of a die are all other outcomes (1, 2, 3, 5 and 6) of the roll. A joint event is an event that has two or more characteristics, such as head on a coin toss and a 4 on the roll of a die. A joint event in forensic context could be both—evidence is observed, and the assertion is true. *Marginal probability* refers to the probability of occurrence of simple events, and *joint probability* refers to the probability of occurrence of two or more events.

Conditional probability refers to the probability of an event, given information about the occurrence of another event. Mathematically, the conditional probability is the joint probability of the two events divided by the marginal probability of the event that is known to have occurred.

Table 6.1 Mathematical Notations and Formulae

Probability Term	Mathematical Notation	Mathematical Formula	Conditions
Marginal Probability of A	$P(A)$		
Conditional Probability of B given A	$P(B|A)$		
Joint Probability of A and B	$P(A \cap B)$	$P(B|A) \cdot P(A)$	
		$P(B) \cdot P(A)$	A and B are independent
Probability of A or B	$P(A \cup B)$	$P(A) + P(B) - P(A \cap B)$	
		$P(A) + P(B)$	A and B are mutually exclusive
Posterior Probability	$P(A|B)$	$\dfrac{P(A \cap B)}{P(B)}$	

Joint probabilities, also known as *event probabilities*, are the relative frequencies of the occurrence of the events. Such probabilities are often termed as unconditional probabilities, as no special conditions are assumed. Often, however, one may have additional knowledge that might affect the outcome of an event or the veracity of an assertion. A probability that reflects such additional information or knowledge is called the *conditional probability*. Graphically through a Venn diagram, the way to compute this conditional probability is to divide the probability of the part of A that falls within the reduced sample space, namely P(B). The conditional probability of an assertion A, given an observation B, is denoted as P(A|B) and is computed by dividing the probability that both A and B occur P(A and B) by the marginal probability of B or P(B). The intuitive reasoning underlying this formula is that the joint probability of A and B is adjusted from its original value in the complete sample space to the reduced sample space P(B).

Consider a medical example in the following charts on probability of developing cancer for smokers and nonsmokers. There are two possible types of individuals, smokers and nonsmokers. Further, there are two possible outcomes: cancer or no-cancer. Note that an individual is either a smoker or a nonsmoker; hence, the outcomes are mutually exclusive. The entire population can be divided into smokers and nonsmokers, thus these are also collectively exhaustive. Similarly, an individual either has cancer or doesn't; hence, these outcomes are also mutually exclusive and collectively exhaustive. As shown in Table 6.2, each cell in the table gives the joint probability of the two corresponding events. For example, the top left-hand cell denotes the joint probability of a smoker with cancer is 0.06. Similarly, the 0.73 in the bottom right cell implies the joint probability of an individual who doesn't have cancer and is a nonsmoker is 0.73.

Table 6.2   Probability of Smoking and Developing Cancer: Joint Probabilities

	Develops Cancer	
Smoker	Yes	No
Yes	0.06	0.19
No	0.02	0.73

Adding up the cells along the row and the column gives us the marginal probabilities of each outcome as shown in Table 6.3:

Table 6.3 Probability of Smoking and Developing Cancer: Marginal Probabilities

	Develops Cancer		Probability of Smoking
Smoker	Yes	No	
Yes	0.06	0.19	0.25
No	0.02	0.73	0.75
Probability of Cancer	0.08	0.92	

As this table shows, the probability of a person with cancer in the entire population is 0.08, or there is a 0.92 probability that an individual doesn't have cancer. Similarly, the probability of smokers in the population is 0.25 and that of nonsmokers is 0.75. The marginal probabilities are obtained by summing the joint probabilities along each row and each column. The sum, on the margin, is the marginal probability of the event or outcome.

The conditional probabilities are obtained by dividing the joint probability in the cell with the respective marginal probability. Thus, there are two ways of obtaining the conditional, dividing with the marginal in the far right column, or the marginal in the bottom row. The conditional probability of A given B is expressed as P(A|B). The second item in the formulation is the attribute whose marginal probability is in the denominator. For example, if we divide the joint probability in the top left cell with the marginal in the far right column, we get P(Cancer|Smoker) as 0.06/0.25 or 0.24. This implies that there is a 0.24 probability that a smoker would develop cancer. On the other hand, if the joint probability in the top left cell is divided by the marginal probability in the bottom row, we get P(Smoker|Cancer) as 0.06/0.08, or 0.75. This implies that given that a person has cancer, there is a 0.75 probability that the person is also a smoker. The conditional probabilities are often called *likelihoods*. In this example, as illustrated in Table 6.4, there are eight possible conditional probabilities or likelihoods.

Table 6.4 Probability of Smoking and Developing Cancer: Conditional Probabilities

Items	Equation	Numerical Value	Conditional Probabilities
P(Cancer| Smoking)	$\dfrac{P(Cancer \cap Smoking)}{P(Smoking)}$	$\dfrac{0.06}{0.25}$	0.24
P(No Cancer| Smoking)	$\dfrac{P(No\ Cancer \cap Smoking)}{P(Smoking)}$	$\dfrac{0.19}{0.25}$	0.76
P(Cancer| No Smoking)	$\dfrac{P(Cancer \cap No\ Smoking)}{P(No\ Smoking)}$	$\dfrac{0.02}{0.75}$	0.03

Items	Equation	Numerical Value	Conditional Probabilities
P(No Cancer| No Smoking)	$\dfrac{P(\text{No Cancer} \cap \text{No Smoking})}{P(\text{No Smoking})}$	$\dfrac{0.73}{0.75}$	0.97
P(Smoking| Cancer)	$\dfrac{P(\text{Cancer} \cap \text{Smoking})}{P(\text{Cancer})}$	$\dfrac{0.06}{0.08}$	0.75
P(No Smoking| Cancer)	$\dfrac{P(\text{Cancer} \cap \text{No Smoking})}{P(\text{Cancer})}$	$\dfrac{0.02}{0.08}$	0.25
P(Smoking| No Cancer)	$\dfrac{P(\text{No Cancer} \cap \text{Smoking})}{P(\text{No Cancer})}$	$\dfrac{0.19}{0.92}$	0.21
P(No Smoking| No Cancer)	$\dfrac{P(\text{No Cancer} \cap \text{No Smoking})}{P(\text{No Cancer})}$	$\dfrac{0.73}{0.92}$	0.79

6.6 The Probative Value of Evidence

The mathematical formulation of evidence, its nature and its strength, is of critical importance in applying statistical techniques to forensic accounting. An understanding of these concepts is essential as it is the foundation of statistical reasoning. In this section, the concepts of probability just discussed are synthesized to address the problems in forensic accounting and fraud detection.

A forensic accountant or a fraud examiner seeks to establish the presence or absence of fraud, leading to the two complementary assertions, A and ~A: existence of fraud and irregularities; absence of fraud and irregularities, respectively. In investigating these assertions, the auditor accumulates evidence from a data trail and other forensic evidence. Although the evidence denoted as E may be qualitative in nature, application of statistical techniques mandates quantification. One such quantification of evidence, based on conditional probabilities, is known as the likelihood ratio and symbolized as λ. The likelihood ratio, λ, is defined as

$$\lambda = \frac{P(E|A)}{P(E|\sim A)}$$

Evidence that is supportive of the assertion is defined by a likelihood ratio of greater than one, or $\lambda > 1$. Evidence conflicting with the assertion is defined by a likelihood ratio between zero and one, or $0 < \lambda < 1$. Evidence having no bearing on the assertion, or irrelevant evidence, is denoted by λ=1. The stronger the evidence is in support of an assertion, the higher the likelihood ratio. Similarly, evidence that strongly refutes the assertion is denoted by a likelihood ratio closer to zero.

An example of joint and conditional probabilities in a business setting is based on the work of Prof. Ariely of Duke University, who researches business ethics and the related topic of honesty among business executives. One interesting hypotheses that he and his students have studied is whether executives who cheat in golf also lie in business dealings. Suppose a joint probability table has been enumerated (see Table 6.5).

Table 6.5 Probability of Cheating in Golf and Lying in Business: Joint Probabilities

	Cheats in Golf	
Lies in Business	Yes (E)	No (~E)
Yes (A)	0.56	0.04
No (~A)	0.24	0.16

Adding up the cells along the row and the column gives us the marginal probabilities of each outcome as shown in Table 6.6.

Table 6.6 Probability of Cheating in Golf and Lying in Business: Marginal Probabilities

	Cheats in Golf		Probability of Lying
Lies in Business	Yes (E)	No (~E)	
Yes (A)	0.56	0.04	0.60
No (~A)	0.24	0.16	0.40
Probability of Cheating in Golf	0.80	0.20	

As before, the marginal probabilities are computed by adding the joint probabilities across the rows or down the columns. Note that the marginal probabilities add up to one. The probability of lying in business is 0.4 and that of not lying is 0.6. Similarly, the probability of cheating in golf is 0.8 and that of not cheating is 0.2.

The conditional probabilities can be obtained by dividing the joint probability in each cell by the corresponding marginal probabilities. Using the equation presented in the previous section, the P(cheats in golf | lies in business), equals 0.56/0.60 or 0.933. Similarly, the conditional probability, P (cheats in golf | does not lie in business), equals 0.24/0.40, or 0.60. That is, an executive who cheats in golf is about 1.5 times more likely to lie in business dealings than an executive who doesn't cheat in golf. Using equations this is computed as follows:

$$\lambda = \frac{P(\text{cheats in golf} \mid \text{lies in business})}{P(\text{cheats in golf} \mid \text{doesn't lie in business})}$$

$$\lambda = \frac{(0.56/0.6)}{(0.24/0.4)} \text{ or } \frac{0.933}{0.6} = 1.555$$

The likelihood ratio, based on conditional probabilities, and Bayes' Rule can be used to compute the change in beliefs in light of new evidence. In particular, the Bayesian approach facilitates the calculation of revised probabilities, when the probabilities are expressed as odds. The joint probability of two events A and B, $P(A \cap B)$ can be written as

$P(A \cap B) = P(B) P(A|B) = P(A) P(B|A)$

Similarly, the joint probability of ~A and B is

$P(B) P(\sim A|B) = P(\sim A) P(B|\sim A)$

Dividing the first equation by the second and canceling P(B):

$$\frac{P(A|B)}{P(\sim A|B)} = \frac{P(B|A)}{P(B|\sim A)} \times \frac{P(A)}{P(\sim A)}$$

This equation, known as the "odds-likelihood" form of the Bayes' Rule, has three components: the posterior odds, the likelihood ratio, and the prior odds. The term on the left side of the equation is the posterior odds, or odds of the event A being true having observed event B. Written in this form it could be expressed simply as the posterior odds become the prior odds for the second item of evidence and so on. Thus, the likelihood ratio has a multiplicative property. That is, the combined strength of two or more items of evidence is the product of the individual strengths of evidence.

This property of likelihood ratio has great significance in evaluating forensic evidence. For independent items of evidence, this representation facilitates assessing the cumulative significance of a variety of evidence

Posterior Odds = Prior Odds × Likelihood ratio

This expression linking the posterior and prior odds has a further important consequence when several items of evidence are aggregated. Given two items of evidence, their combined strength is the product of the individual strengths. In forensic accounting and auditing usually there are multiple items of evidence bearing on the assertions, the multiplicative property of this formulation eases the representation and combination of such multiple items of evidence. This property is illustrated at the end of the next section after a discussion on the Bayes' theorem.

6.7 Bayes' Rule

Conditional probabilities facilitate the transformation of information into evidence. For instance, medical researchers try to diagnose the disease based on observed symptoms. An investor decides to purchase stock based on a favorable rating from a financial analyst. In forensic accounting and auditing, the use of conditional probabilities is extremely pertinent, as the forensic accountant attempts to assess the occurrence of fraud and other financial irregularities based on observation of financial data. However, in many practical situations while the occurrence of the event, or evidence, is known with certainty, the chances of the event occurring before the fact are difficult to assess directly. That is, in the equation of conditional probabilities

$$P(A|E) = \frac{P(A \cap E)}{P(E)}$$

the denominator $P(E)$ is usually unknown. For these types of problems, an extension of conditional probabilities through Bayes' Rule (also known as Bayes' theorem) has to be applied.

Reverend Bayes (1702–1761) was a religious minister and a mathematician. As a mathematician, he was fascinated by games of chance and rules of probabilities. As a minister, he was passionate about proving the existence of God. Consequently, he set out to prove the existence of God by examining and interpreting the sample evidence of the world around him. The mathematicians prior to his time were focused on predicting the results of games of chance. That is, they were interested in predicting future events given the causes in the present or the past. Reverend Bayes was interested in the inverse problem of ascertaining the probability of "causes" based on the observed "events." He wanted to draw conclusions on the hypotheses regarding the existence of God from the observations of consequences. His efforts led to a wide body of knowledge known as Bayesian Decision Theory, developed extensively in the mid-twentieth century. This theorem, which was published posthumously, bears his name and has been the hallmark of decision-making under uncertainty. Bayes' theorem forms the basis for revising prior probabilities in light of new evidence.

Bayes' Rule is best developed through an example. Suppose a company has three accountants, A, B and C, in charge of data entry. The error rates for these individuals vary. While

A's error rate is 0.01, B's is 0.03 and C's is 0.07. On a particular day, A entered 60% of the sales data, while B entered 30% of sales data, and C entered the remaining 10%. At the end of the day the supervisor notices that the amount of total sales recorded did not match with the total entered by the three operators. The supervisor wants to determine the order in which the three operators should verify their work. As A did most of the data entry for the day, should A be the first to check his or her work? The problem can be solved using conditional probabilities and the Bayes' Rule as follows:

$$P(A \text{ process the transaction} \mid \text{error was detected}) = \frac{P(A \text{ processed and error})}{P(\text{error})}$$

$$P(B \text{ process the transaction} \mid \text{error was detected}) = \frac{P(B \text{ processed and error})}{P(\text{error})}$$

and

$$P(C \text{ process the transaction} \mid \text{error was detected}) = \frac{P(C \text{ processed and error})}{P(\text{error})}$$

Although the manager does not know any of these probabilities directly, he knows the following

P(A processed a randomly selected transaction) = 0.6,
P(error| A has processed) = 0.01

P(B processed a random transaction) = 0.3,
P(error| B has processed) = 0.03

P(C processed a random transaction) = 0.1.
P(error| C has processed) = 0.07

The only information that the manager needs to be able to compute these conditionals is the denominator, or P(error) because using the six probabilities here, he can derive the numerator of the three equations as

P(A processed and error occurs) = P(A processed a random transaction) × P(error| A processed)

= (0.6) (0.01)

= 0.006

P(B processed and error) = P(B processed a random transaction) × P(error| B processed)

= (0.3) (0.03)

= 0.009

P(C processed and error) = P(C processed a random transaction) × P(error| C processed)

$\quad$ = (0.1) (0.07)

$\quad$ = 0.007

Thus, the a-priori probability of error = P(A processed and error) + P(B processed and error) + P(C processed and error).

$\quad$ = 0.006 + 0.009 + 0.007

$\quad$ = 0.022

Now that we know that the probability of error, P(E) is 0.022, we can substitute this as the denominator in the given equations and solve for the posterior probabilities as follows

$$P(A \text{ processed} | \text{error}) = \frac{P(A \text{ processed and error})}{P(\text{error})} = \frac{0.006}{0.002}$$

P(A processed | error) = 0.2727

$$P(B \text{ processed} | \text{error}) = \frac{P(B \text{ processed and error})}{P(\text{error})} = \frac{0.009}{0.022}$$

P(B processed | error) = 0.4091

$$P(C \text{ processed} | \text{error}) = \frac{P(C \text{ processed and error})}{P(\text{error})} = \frac{0.007}{0.002}$$

P(C processed | error) = 0.3182

In this example, though A processed the highest number of transactions, it is more likely that B processed the transaction in error. In fact, it is least likely that A processed the transaction in error because A has an error rate so much less than the other two accountants. Though A processed more transactions, it is unlikely that he or she processed the transaction in error.

We can now mathematically define the Bayes' Rule.

$\quad$ Let A_i, (i = 1, 2, 3, ..., n) be a set of mutually exclusive and collectively exhaustive events.

$\quad$ Let B be another event which is preceded by an A_i event.

$\quad$ And $P(A_i)$ and $P(B|A_i)$ are known.

Then $P(A_1|B) = \dfrac{P(A_1)P(B|A_1)}{\sum_{i=1}^{n} P(A_i)P(B|A_i)}$

Now to continue with the example in the previous section to illustrate the mathematics of combining two items of independent evidence: Recall the example was about the conditional probabilities of an executive lying in business (A) and cheating in golf (E). We had computed the following conditional probabilities in Section 6.6: P(Cheats in Golf | Lies in Business) = $P(E|A)$ = 0.933 and P(Cheats in Golf | Does not lie in business) = $P(E|\sim A)$ = 0.6. Now consider a second related event of routinely violating the speed limit and the premise that executives who routinely violate speed limits are also prone to lie in business. The joint probability distribution is enumerated in Table 6.7.

Table 6.7 Probability of Routinely Violating Speed Limits and Lying in Business: Marginal Probabilities

	Violates Speed Limits		Probability of Lying
Lies in Business	Yes (E_2)	No ($\sim E_2$)	
Yes (A)	0.42	0.18	0.60
No ($\sim$A)	0.14	0.26	0.40
Probability of Violating Speed Limits	0.56	0.44	

As before, the marginal probabilities are computed by adding the joint probabilities across the rows or down the columns. Note that the marginal probabilities add up to one. The probability of lying in business, P(A) is 0.6 and that of not lying P($\sim$A) is 0.4, same as before. Similarly, the probability of violating the speed limit is 0.56 and that of not violating the speed limit is 0.44.

The conditional probabilities can be obtained by dividing the joint probability in each cell by the corresponding marginal probabilities. Using the equation presented for the previous section, the P(violates speed limit | lies in business), or $P(E_2|A)$ equals 0.42/0.60 or 0.70. Similarly, the conditional probability, P (violates speed limit | does not lie in business), or $P(E_2|\sim A)$ equals 0.14/0.40, or 0.35. Now if one knows that an executive routinely violates speed limits, this person is about twice as likely to lie in business relative to an executive who adheres to the speed limits. Using equations this is computed as follows:

$$\lambda_2 = \dfrac{P(\text{violates speed limit} \mid \text{lies in business})}{P(\text{violates speed limit} \mid \text{doesn't lie in business})} = \dfrac{0.42/0.6}{0.14/0.4} = \dfrac{0.7}{0.35} = 2.0$$

Now suppose there is an executive we know nothing about. Hence, we set the prior probability that he lies in business to be 0.6, equal to the proportion of managers in the overall population that lie, or a prior odds of 1.5.[1] Now we observe this executive cheating in golf (E). Based on this evidence we revise our prior beliefs in accordance with Bayes' Rule as follows

$$P(A|E) = \frac{P(E|A)P(A)}{P(E|A)P(A) + P(E|\sim A)P(\sim A)}$$

$$P(A|E) = \frac{(0.933)(0.6)}{(0.933)(0.6) + (0.6)(0.4)}$$

$$P(A|E) = \frac{0.56}{0.8} = 0.7$$

$P(A|E)$ is known as the posterior probability, and in this example equals 0.7 once it is observed that the executive cheats in golf. The posterior odds can be computed by dividing 0.7 with the posterior probability $P(\sim A|E)$, which equals 0.3. Thus, posterior odds equal 2.3333. The posterior odds could alternatively be obtained by multiplying the prior odds of 1.5 with the strength of evidence of 1.555555, which also equals 2.333.

Now suppose it is observed that the executive routinely violates speed limits as the second item of evidence, or E_2. Based on this second item of evidence we again revise our beliefs in accordance to Bayes' Rule as follows

$$P(A|E \& E_2) = \frac{P(E_2|A)P(A)}{P(E_2|A)P(A) + P(E_2|\sim A)P(\sim A)}$$

$$P(A|E \& E_2) = \frac{(0.7)(0.7)}{(0.7)(0.7) + (0.35)(0.3)}$$

$$P(A|E \& E_2) = \frac{0.49}{0.595} = 0.8235$$

$P(A|E \& E_2)$ is known as the posterior probability of the executive lying in business given two items of evidence: cheats in golf; and routinely violates speed limits. In this example the posterior probability equals 0.8235 once it is observed that the executive cheats in golf and routinely violates speed limits. The posterior odds can be computed by dividing

0.8235 with the posterior probability $P(\sim A|E)$ which equals 0.1765. Thus, posterior odds equal 4.6667. This could alternatively be obtained by multiplying the prior odds of 1.5 with the combined strength of evidence. The strength of evidence of cheating in golf was 1.5555 and that of violating speed limits was 2.0. The combined strength of evidence is obtained by taking the product of the two strengths of evidence, which is 1.5555×2.0, equals 3.1111. Multiplying this combined strength of evidence of 3.111 with the prior odds of 1.5, we obtain the posterior odds as 4.6667, identical to the amount obtained by applying Bayes' Rule.

Bayes' Rule is applied in situations where new information changes the probability of the events of interest. These changes in probability are called *belief revisions*. Belief revisions have wide applicability in forensic accounting as new items of evidence collected sequentially prompts auditors to reassess their beliefs on the assertions.

6.8 Chapter Summary

This chapter presented the basic notions of probability, the definitions, and mathematical notations. Although marginal probabilities and joint probabilities are usually emphasized in business statistics, the concept of conditional probability is pertinent to auditing and forensic accounting. The Venn diagrams presented earlier in the chapter help illustrate important relationships between evidence and assertion. As evidence is accumulated in a forensic accounting case, it may be helpful to visualize the items of evidence as a Venn diagram. Also it is critical to distinguish between the situations when evidence confirms an assertion and when an assertion always causes the evidence. Although observation of evidence in the former case makes the assertion a certainty, the observation of evidence in the second case is not as compelling.

The intuitions obtained from the Venn diagrams were formalized in mathematical notations through conditional probabilities. The mathematical formulation and distinctions among prior, marginal, joint, conditional, and posterior probabilities were presented. These definitions and concepts are fundamental to the disciplines of probability and statistics. Thus, an understanding of this terminology is important for the forensic accountant to be able to communicate with statisticians, legal professionals, and other experts outside of accounting. These concepts were illustrated through the use of numerical examples. Although being able to reproduce the mechanics of these computations may not be critical for practicing forensic accountants, an understanding of the computations will enable them to identify situations in which these concepts have not been properly applied.

The concept of strength of evidence in terms of likelihood ratio was introduced in this chapter. Though theoretical and normative in nature, the concepts provide a necessary structure and form the foundation of evidence evaluation. Though difficult to apply in practice due to the unavailability of conditional probabilities, these notions provide the basis for statistical reasoning. A grasp of these concepts will further the understanding of more sophisticated statistical concepts that follow.

This chapter concluded by presenting the widely applied Bayes' Rule. Bayes' Rule, formulated in the mid-eighteenth century, reverses probabilistic reasoning from determining probabilities of what effects would be caused to that of assessing what caused those effects. The difference is not just semantics, in the former the cause is known for certain; whereas in the latter the effect has been observed with certainty. In a forensic accounting context the latter is more relevant. In these settings, the accountant observes the evidence and assesses the possibility that a particular action (such as fraud or financial impropriety) was the underlying cause. In the subsequent chapters, the concepts presented in this chapter, particularly Bayes' Rule, will be helpful in relating established statistical techniques to forensic accounting.

6.9 Endnote

1. Probability of an assertion divided by the probability of its negation results in the odds. Thus 0.6/0.4 gives the prior odds of 1.5.

7

Discrete Probability Distributions

Probability theory is nothing but common sense reduced to calculation.
—Pierre Simon Laplace, *Théorie Analytique des Probabilitiés*, 1812

7.1 Introduction

The previous chapter presented the basic concepts of probability and mathematical notation. In this chapter the connections between descriptive statistics and probability distributions are established. In a forensic accounting context, determining the appropriate data-gathering procedures, the appropriate intervals, and the right methods of presenting the data could all be complex judgments and require the use of statistical planning tools. Fortunately, many of the data characteristics though seemingly unrelated share similar underlying properties that enable the use of certain probability distributions. When the data characteristics can effectively be captured by a theoretical probability distribution, the forensic accountant can take advantage of statistical results that have already been established. Thus, it is imperative for forensic accountants to become comfortable with commonly used probability distributions if they are to apply these effectively in their investigations. Not only should the forensic accountant be knowledgeable about the application but more importantly about the defining characteristics of specific distributions.

Probability distributions, like frequency distributions, describe how outcomes may be expected to vary. The concept of probability distribution is central in statistical inference and in managing uncertainty. Probability distributions are classified as either discrete or continuous depending on the nature of the variable under consideration. A discrete random variable is one that can only take on distinct values, but a variable is continuous if it takes on any value in a given range. This chapter discusses the discrete probability distribution, and in the following chapter this discussion is extended to include continuous probability distributions.

Three common discrete probability distributions need to be considered specifically: the binomial distribution, the Poisson distribution, and the hypergeometric distribution. All of these distributions describe uncertain situations with discrete values for the variable of interest. Many accounting and auditing data-sets can be approximated by one of these distributions. This chapter also covers how discrete probability distributions are developed and how they indicate the type of events they describe. We further discuss several descriptive measures that help define these distributions.

7.2   Generic Definitions and Notations

An experiment, trial, or query results in some form of outcome or event. When the trial has a quantitative characteristic, like the amount of error, we can associate a number with each outcome. Say an auditor examines 15 randomly selected accounts that form the experiment or trial. The outcome of this experiment is the number of accounts that had inaccurate balances. This number could be represented by the variable X. In this situation X is a random variable with the following values: 0, 1, 2, 3, ..., 14, 15. The random variable takes on only discrete value; that is, the accountant is not going to find 4.5 errors as a result of his investigation. A graph that plots the number of errors on repeated trials of 15 accounts forms the basis of probability distribution. However, the graphs are of limited usefulness as only partial information can be described in a graph. Often knowledge of mean and standard deviation is of importance. The mean measures the central tendency; the standard deviation measures the dispersion or spread of the probability distribution.

The mean of the probability distribution is also known as the expected value of the random variable. It is measured as a weighted average of the values of the random variables weighted by the respective probability assigned to those values. More formally, if X is a discrete random variable that takes on the value of x with probability P(x), then the expected value of X, denoted as E(X) or as the mean µ, is

$$E(X) = \mu = \sum x \times P(x)$$

where: E(X) is the expected value of the random variable X

µ is the mean of the discrete probability distribution

x are the discrete values of the random variable

P(x) is the probability of each value of x

As can be seen, the expected value of the random variable is the weighted average of the variable's possible values.

The standard deviation measures the spread or the dispersion of the data around the mean. Low values of standard deviation imply that the random variable is narrowly dispersed around the mean. Higher values of standard deviation imply that the random variable is widely dispersed around the mean. The standard deviation is the square root of the weighted average of squared differences between each value of the random variable and the expected value. Mathematically,

$$\sigma = \sqrt{\sum [X - E(X)]^2 \, P(X)}$$

Table 7.1 illustrates these equations through a simple example. Suppose the probability distribution of clients calling the company each day to report errors in billing is as follows:

Table 7.1   Computation of Mean and Standard Deviation

x	P(x)	x.P(x)	$(x-\mu)^2$	$(x-\mu)^2 \times P(x)$
10	0.08	0.8	210.25	16.82
15	0.12	1.8	90.25	10.83
20	0.25	5.0	20.25	5.0625
25	0.20	5.0	0.25	0.05
30	0.15	4.5	30.25	4.5375
35	0.12	4.2	110.25	13.23
40	0.08	3.2	240.25	19.22
Total	1.00	24.5		69.75

The expected value or mean is 24.5 calls per day, and the standard deviation is the square root of 69.75 or 8.35 calls.

The appendix illustrates these computations by using an Excel worksheet.

7.3   The Binomial Distribution

The simplest probability distribution describes experiments with only two outcomes, as in the case of a coin toss. Many accounting events can be modeled through this process. When an auditor inspects inventory items for errors there are only two outcomes possible, error or no error. Similarly, in evaluating proper functioning of internal controls, the auditor might check for a supervisor's signature. Again, only two outcomes are possible—either the signature is present or absent. A process in which each trial or observation can have only one of two states is called a *binomial* or a *Bernoulli process*.

Chapter 5, "Data Mining," used the Bernoulli vector in the context of the shopping cart analogy to appreciate the use of association analysis.

A binomial distribution can be described to have the following characteristics:

- The process has only two possible outcomes: success and failure, error or no error, head-tail, and so on.
- Each and every trial or experiment is identical.
- The trials are independent of each other. The outcome of one does not in any way influence the outcome of another trial. Finding an error in one account does not alter the chances of finding error in another.
- The process is stationary in that the probability of success remains constant from trial to trial.

Suppose an auditor is testing the cash disbursement system and verifies whether the amount paid to a claimant equals the amount on the invoice. If it does, it is labeled correct, else an error. For simplicity, assume that the error rate is 5%, or the probability of error is 0.05 and that of success or no error is 0.95. Further, the disbursement system is such that the Bernoulli process applies, and the following conditions are true:

- There are only two possible outcomes when a payment is made: Either it is correct or it is in error.
- Each payment is made through the same cash disbursement system.
- The outcome of any paid invoice is independent of whether the preceding invoices had errors or not.
- The probability of error remains constant at 0.05, or $P(e) = 0.05$.
- The probability of correct is 0.95, or $P(c) = 0.95$.

Suppose the auditor samples four disbursements to be tested for errors. As the sample is picked at random, the number of errors will vary from sample to sample. The sample outcome, measured in terms of incidence of error is a discrete variable, say X. Variable X can take five distinct values: 0, 1, 2, 3, and 4. As the outcome of each item in the sample is independent of others, the probability of each value of X can be obtained by using the multiplicative rule. That is, the probability of 0 errors in a sample of 4 can be obtained as

$$P(X = 0) = P(c, c, c, c) = P(c) \times P(c) \times P(c) \times P(c)$$

Since P(success) = 0.95

$$P(X = 0) = (0.95)^4$$

Or $P(X = 0) = 0.8145$

Similarly, the probability of finding exactly one error in the sample of 4 can be found by using the multiplicative rule as well as the addition rule for mutually exclusive events.

$$P(X = 1) = P(e, c, c, c) + P(c, e, c, c) + P(c, c, e, c) + P(c, c, c, e)$$

$$P(e, c, c, c) = P(e) \times P(c) \times P(c) \times P(c)$$

Or (0.05) (0.95) (0.95) (0.95)

Likewise P (c, e, c, c) = (0.95) (0.05) (0.95) (0.95),

P (c, c, e, c) = (0.95) (0.95) (0.05) (0.95), and

P (c, c, c, e) = (0.95) (0.95) (0.95) (0.05).

Then $P(1 \text{ error}) = 4 (0.95)^3 (0.05)$

= 0.1715.

The probability of finding exactly two errors in the sample of 4 can be similarly found by using the multiplicative rule as well as the addition rule for mutually exclusive events.

$$P(X = 2) = P(e, e, c, c) + P(e, c, c, e) + P(e, c, e, c) + P(c, e, e, c) + P(c, e, c, e) + P(c, c, e, e)$$

$$P(e, e, c, c) = P(e) \times P(e) \times P(c) \times P(c)$$

Or (0.05) (0.05) (0.95) (0.95)

Likewise P (e, c, e, c) = (0.05) (0.95) (0.05) (0.95)

P (e, c, c, e) = (0.05) (0.95) (0.95) (0.05)

P (c, e, e, c) = (0.95) (0.05) (0.05) (0.95)

P (c, e, c, e) = (0.95) (0.05) (0.95) (0.05) and

P (c, c, e, e) = (0.95) (0.95) (0.05) (0.05)

Then $P(2 \text{ errors}) = 6 (0.95)^2 (0.05)^2$

= 0.0135

The probability of finding exactly three errors in the sample of 4 can be similarly found.

$$P(X = 3) = P(e, e, e, c) + P(e, e, c, e) + P(e, c, e, e) + P(c, e, e, e)$$

$$P(e, e, e, c) = P(e) \times P(e) \times P(e) \times P(c)$$

Or (0.05) (0.05) (0.05) (0.95)

Likewise P (e, e, c, e) = (0.05) (0.05) (0.95) (0.05)

Then P (e, c, e, e) = (0.05) (0.95) (0.05) (0.05) and
P (c, e, e, e) = (0.95) (0.05) (0.05) (0.05)
P(3 errors) = 4 (0.05)³ (0.95)
= 0.0005

Finally, the probability of finding all four items to be in error is

P(X = 4) = P(e, e, e, e)

P(e, e, e, e) = (0.05)⁴ = 0.0000062

Note that the sum of probabilities of finding 0, 1, 2, 3 or 4 errors add to 1. That is, 0.8145 + 0.1715 + 0.0135 + 0.0005 equals one. Each pair of values n and p establishes a different binomial distribution. Thus, the binomial is in fact a family of probability distributions. Computations are laborious for large values of n and Excel macros have been developed to ease computations. Also tables are available for determining cumulative probabilities for binomial distribution.

For small sample sizes, the binomial distribution can be generated by enumerating the sample space and proceeding with the computations as just shown. However, as the sample size increases and becomes large and the different events and possibilities increase, this method becomes tedious. For such situations, instead use the binomial formula shown here:

$$P(X_1) = \frac{n!}{X_1! X_2!} p^{X_1}(1-p)^{X_1}$$

Where: n is the sample size

X_1 is the number of errors

n! is read as 'n factorial' and is the product of n, n-1, ...3, 2, 1.

X_2 is the number of non-errors or (n − X_1), and

p is the probability of error.

The mean and standard deviation for the binomial distribution can be found as before. Recall from the previous section that the mean or the expected value of a discrete random variable X can be computed as

$$E(X) = \sum X P(X)$$

where: E(X) is the expected value of X

X assumes all possible values of the random variable

P(X) is the probability for a specific value of X

Using the formula just shown in the previous example, the expected value of errors in a sample of 4 can be found as shown in Table 7.2.

Table 7.2 Expected Value Computation

Number of Errors (X)	P(X)	X × P(X)
0	0.8145	0
1	0.1715	0.1715
2	0.0135	0.0270
3	0.0005	0.0015
4	0.0000	0
Total	1.000	0.2000

Therefore if the probability of error in any one payment is 0.05, the average number of errors found in a repeated sample size of four is equal to 0.20. Of course, for any single sample one would not find 0.200 errors because the number of errors must occur in discrete values of 0, 1, 2, 3, or 4. In general, for binomial distribution the mean can be found through an easier formula such as

$$\text{Mean} = \mu = np$$

where, n is the sample size and p is the probability of error

In this example, n is equal to 4, and p is equal to 0.05. Thus, the mean is 4 times 0.05 or 0.2 as obtained previously.

Recall from the previous section that the standard deviation of a discrete probability distribution is given as

$$\sigma = \sqrt{\sum [X - E(X)]^2 \, P(X)}$$

where, X is the specific value of the random variable

σ is the standard deviation of X

E(X) is the expected value or mean of the random variable

P(X) is the probability of X

Continuing with the previous example for a sample of 4 and probability of error of 0.05, the standard deviation for the distribution is shown in Table 7.3.

Table 7.3   Computation of Standard Deviation

X	P(X)	X − E(X)	[X − E(X)]²	[X − E(X)]² P(X)
0	0.8145	−0.2	0.04	0.03258
1	0.1715	0.8	0.64	0.10976
2	0.0135	1.8	3.24	0.04374
3	0.0005	2.8	7.84	0.00392
4	0.0000	3.8	14.44	0.0000
Total				0.19

The standard deviation can be found by taking the square root of 0.19 as 0.4359. Again, instead of undertaking the detailed computation as just described, for binomial distributions the standard deviation is defined as

$$\sigma = \sqrt{np(1-p)}$$

where, σ is the standard deviation,
n is the sample size,
p is the probability of finding an error, and
(1-p) is the probability of not finding an error.

For this example, n is 4, and p is 0.05; therefore, (1 − p) is 0.95, and by substituting in the formula the standard deviation σ is 0.4359 as before. Thus the mean and standard deviation for this distribution are 0.200 and 0.4359, respectively.

In a binomial distribution if the probability of success is 0.5, as in a coin toss, the binomial distribution is symmetrical regardless of the sample size. This is illustrated in Figure 7.1 which shows frequency distributions for samples of 10 and 100. Both the distributions are centered at the expected value or the mean and are symmetrical around it. The red line drawn on each histogram resembles the bell curve of a normal distribution that is introduced in the following chapter. As the probability differs from 0.5 in either direction, the binomial distribution becomes skewed. With the increase in sample size, the skewed nature of it is reduced, and the distribution tends to become symmetrical. The asymmetry however is most pronounced in situations where p is very low and the sample size is small. This was the situation in the example with p = 0.05 and n = 4, hence the probability distribution was highly skewed. Additional examples are graphed in Figure 7.2. Note that as the sample size increases, the distribution becomes more symmetrical around the mean.

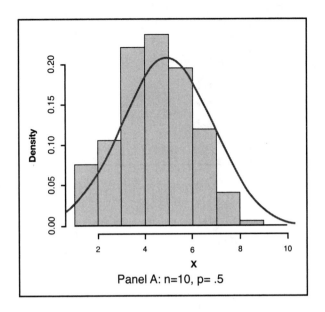

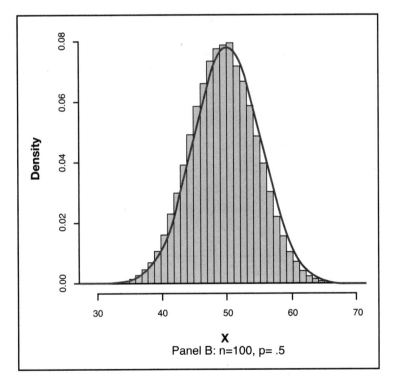

Figure 7.1 Binomial distribution

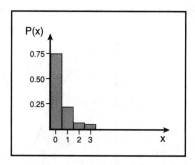

Panel A: n=5; p=0.05

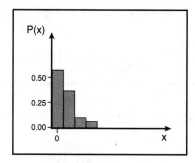

Panel B: n=10; p=0.05

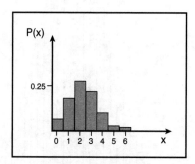

Panel C: n=50; p=0.05

for p = 0.05 and varying sample size

Figure 7.2   Skewed binomial distribution

The symmetry of binomial distribution is an important property, as is discussed in the next chapter, and allows approximation through normal probability distribution. In the plots, the red line denotes the corresponding normal distribution to illustrate the proximity of the binomial probability distribution to the normal probability distribution. In most forensic accounting and other auditing situations the probability of error is small, closer to zero than to 0.5, small sample sizes compound the problems of making statistical inferences on these cases using the results and properties of normal probability distribution. This is an important caveat for probability and statistical applications in forensic accounting and auditing, which could be corrected by having a sufficiently large sample size.

7.4 Poisson Probability Distribution

In certain accounting and auditing situations the number of errors can be ascertained, but the number of non-errors cannot be ascertained. In such situations the total number of outcomes cannot be determined, and hence binomial distribution cannot be effectively applied. This is the case with most continuous auditing applications. In these situations, when the count of only the outcome of interest is known but there is no information on the total number of outcomes, the Poisson distribution can be applied. For instance, Poisson distribution can be used in a continuous auditing situation to determine the number of compromises to internal controls that occurred during a week.

The application of Poisson distribution is appropriate when the following underlying characteristic of data are satisfied:

- The event is considered a rare event. The probability of occurrence of the event should usually be less than 0.1.

- The events must be random and independent of each other. That is, the event occurring must not be predictable, nor must it influence the chances of another event occurring.

The Poisson distribution is described by a single parameter, λ (lambda) which is the average occurrence per segment. The value of λ can be estimated from the occurrences and is context-specific. For example, λ could be the average number of calls for emergency services in a day or the average number of machine breakdowns in a month, or the average number of trucks arriving at a weigh station in an hour. Poisson distribution is a discrete probability distribution with a range that consists of all non-negative integers. If modeling the number of flight arrivals at an airport in a given hour, only the integer values can occur; that is, 1.5 planes cannot land. It is important to note, however, that the mean of the distribution could be a number that is not one of the values of the distribution. That is, it is acceptable to have 1.5 as the mean of the number of planes that land in an hour, even though 1.5 planes could never actually land.

The Poisson probability distribution specifies the probability of a certain number of occurrences over a specified time using the formula

$$P(X=x) = \frac{(\lambda t)^x e^{-\lambda t}}{x!}$$

where, t is the specified time segment expressed in compatible units,

x is the number of occurrences in time segment t

λ is the estimated parameter of the Poisson distribution

e is the natural base equal to 2.71828

x! is the factorial of x, or product of x, (x-1), (x-2), ..., 3, 2, 1

As can be seen from the equation, the Poisson distribution has only one parameter, λ. That is, to specify a Poisson probability distribution only the value of λ has to be estimated. The parameter λ can be interpreted as the arithmetic mean number of occurrences per interval of time or space that characterizes the process of generating the Poisson distribution.

The application of Poisson probability distribution can be illustrated through a simple numerical example of market demand. Product demand is a common business application of Poisson distribution. It can be viewed as a process that produces random occurrences over a continuous time. Suppose a retailer has estimated the demand of a certain item to be 300 units a month (30 days). The manager is interested in knowing the probability that exactly 7 units would be sold in a given day. The parameters of the problem are as follows:

$\lambda = 300$ per month,

$t = 1$ day or $\dfrac{1}{30}$ of a month,

$\lambda t = 300/30 = 10$.

Then using the Poisson distribution with the value x equal to 7, the probability distribution can be written as

$$P(X=7) = \dfrac{(10)^7 e^{-10}}{7!} = 0.0901$$

Cumulative probabilities, such as the P (X < 10) or P(3 ≤ X ≤ 8) can be computed either by adding individual probabilities or through the use of Poisson probability tables. The appendix discusses the use of a scientific calculator or Excel worksheet applications for Poisson probability distributions.

The mean, variance, and standard deviation of a Poisson distribution are

$\mu = \lambda t$

$\sigma^2 = \lambda t \sqrt{\lambda t}$

where, μ is the mean of the distribution,

λ is the distribution parameter,

σ^2 is the variance, and

σ is the standard deviation.

Note that for a Poisson distribution, the variance of the distribution equals the mean of the distribution. Hence, for processes that are assumed to follow a Poisson distribution, variance can be controlled by controlling the mean.

Poisson distribution can be used to approximate a binomial probability function when probability (p) is small and the number of occurrences (n) is large. Recall from the earlier discussion on binomial distribution that it is not symmetric around the mean when the parameter p in binomial distribution is close to zero, instead the binomial probability distribution is skewed. This lack of symmetry limits the applicability of normal probability distribution to those data-sets. In such situations however, it can be mathematically proven that λ can be approximated as n×p, and the Poisson distribution can be applied. In this context, the Poisson distribution gives the probability of observing x occurrences in n trials where p is the probability of occurrence on a single trial. That is, x, p, and n are interpreted in the same way as in binomial distribution. As there are mathematical assumptions underlying the derivation of the Poisson distribution from the binomial distribution, these assumptions require that n is large and p is small. The probability distribution plot for Poisson is shown in Figure 7.3. Again, the red line drawn through the histogram denotes the corresponding normal distribution.

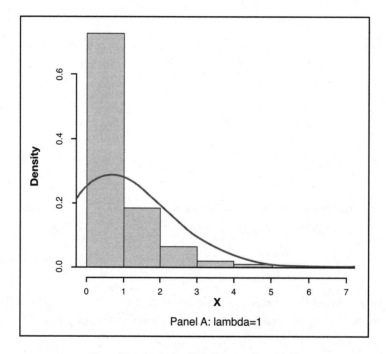

Figure 7.3 Poisson distribution (Panel A)

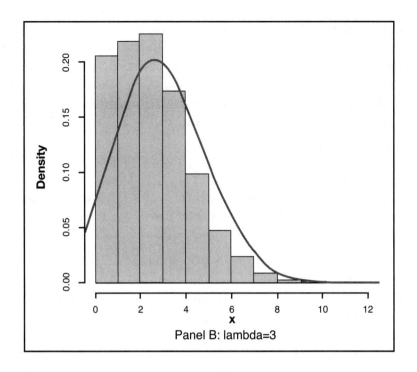

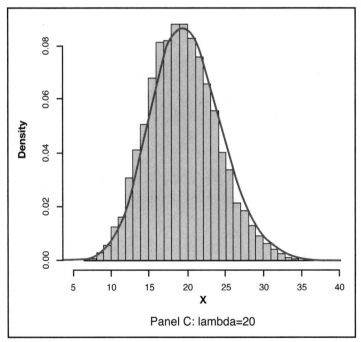

Figure 7.3 Poisson distributions (Panels B and C)

In forensic accounting and auditing, Poisson probability distribution has applications in the area of cyber-security. It has been used to detect anomalies of volume in computer network traffic. These types of anomalies can be caused by an attack or intrusion, a computer virus, a flash crowd, and such. In all cases it is associated with a sudden change (usually rapid increase) in the network traffic's volume characteristics. The network traffic pattern is shown to be well described by a Poisson process.[1] The average network packet rate is modeled as the arrival rate of a Poisson process, which undergoes an upsurge triggered by an anomaly. The average packet rate in network traffic is measured in the thousands, or n is very large. Also, the frequency of upsurge is relatively low, or p is very small. Hence, Poisson distribution can reasonably estimate the process through mean and variance that equal the average packet rate. Large deviations from the mean can be detected sooner.

Poisson probability distribution has also been applied to detect computer hackers entering the system. It helped in the design of an efficient method, which balanced between missing malicious detections and false alarms. Under normal circumstances, the network traffic pattern, which includes the "silent time" between arrivals of consecutive packets is well described by a Poisson process. The malicious attacks on the network, however, do not follow the Poisson process. Hence, when expectations can be derived from the Poisson process, large deviations from those expectation indicate the possibility of malicious attacks.[2]

7.5 Hypergeometric Distribution

Hypergeometric distributions are applicable when certain assumptions of binomial distribution are not supported. The binomial distribution assumes that the items are drawn independently, such that the probability of selecting an item is constant throughout the sampling process. This assumption is easily satisfied when selecting a small sample from a large population. However, in instances when the selection is made from a small population, say hiring decisions, this assumption is violated. For example, if we were to select 3 candidates from a pool of 12 candidates, the probability that someone being selected first is $1/_{12}$ or 0.083, someone selected second is $1/_{11}$ or 0.091, and someone selected third is $1/_{10}$ or 0.10. Thus, the probabilities change with each selection. As the assumptions of binomial distribution are not met, a different probability model has to be developed. The resultant probability distribution is known as the hypergeometric probability distribution.

Both the binomial probability distribution and the hypergeometric probability distribution are concerned with the number of events of interest in a sample consisting of some observations. The approaches vary on how the sample is selected. For the binomial distribution, the sample is selected with replacement from a finite population and with or without replacement from a large population. This makes the probability of selection of any item constant regardless of the order in which it was selected. For the hypergeometric distribution the sample is selected without replacement from a small population. Thus, the selection into sample and the outcome of one observation is dependent on the outcomes of the previous observation. Continuing with this example of hiring from a pool of 12 candidates, assume that 7 of the candidates are male and 5 are female. If the question of interest is whether the second candidate selected is a female, the probability changes based on the outcome of the first selection. If the first selection were a male, the probability of a female being selected second is $5/_{11}$ or 0.4545, however if the first candidate selected were a female, the probability lowers to $4/_{11}$ or 0.3636. Similarly, the probability of a female being selected third depends on the outcomes of the first two selections and are 0.5, 0.4, and 0.3 if none, one, or both candidates selected previously were females, respectively. This dependence of probability on previous outcomes violates an important assumption of binomial distribution.

The probability distribution function of a hypergeometric is derived using the classical definition of probability and the formula for permutations and combinations. Combinatorial probabilities deal with the number of different ways that a certain number of objects can be selected from a small population where the order of selection is not a concern. The number of possible selections is called the combinations and is denoted by $\binom{n}{x}$ where x objects are chosen from a total of n. The numerical value is obtained through the use of factorials, which is the product of n, n-1, n-2, ..., 3, 2, 1 as

$$\binom{n}{x} = \frac{n!}{x!(n-x)!}$$

The hypergeometric probability distribution function of finding x errors in a sample of n, selected from a population of N in which there are s errors is given as

$$P(x) = \frac{\binom{s}{x} \times \binom{N-s}{n-x}}{\binom{N}{n}}$$

where, N is the size of the population
s is the number of errors in the population
n is the sample size
x is the number of errors in the sample

As the number of errors in the sample, x, cannot be greater than the number of errors in the population, s, nor can it be greater than the sample size, n, the range of a hypergeometric distribution is limited to the number of errors in the population or the sample size, whichever is smaller. That is, x satisfies both conditions

x ≤ s, and

x ≤ n.

The mean and standard deviation of the hypergeometric function can be computed as

$$\mu = E(X) = \frac{ns}{N}$$

$$\sigma = \sqrt{\frac{ns(N-s)}{N^2}} \times \sqrt{\frac{(N-N)}{(N-1)}}$$

The probability distribution plot of a hypergeometric distribution is presented in Figure 7.4.

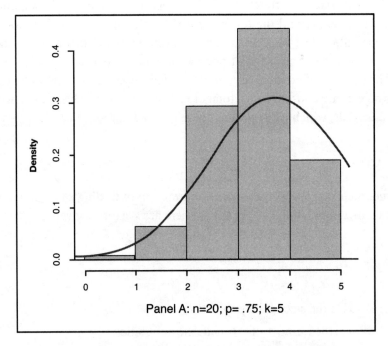

Figure 7.4 Hypergeometric distribution (Panel A)

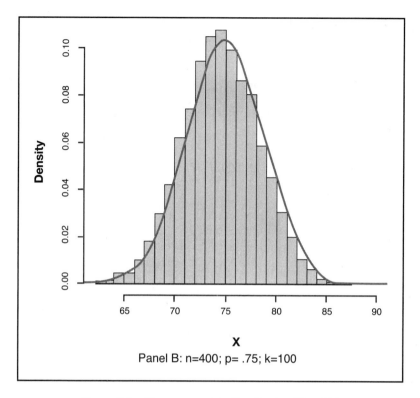

Figure 7.4 Hypergeometric distribution (Panel B)

To illustrate the hypergeometric distribution, consider a multilocation inventory audit problem, a complex problem in financial audits.[3] Suppose a client has 20 inventory locations spread all over the U.S. As visiting each individual location is costly and time consuming, the audit team has a policy of visiting 6 locations each year. The locations to visit are randomly selected. If there are 2 inventory locations that have materially overstated inventory balance, the probability that the auditor will find no locations with errors in his sample can be computed using the equation just given, as follows

$$P(x) = \frac{\binom{s}{x} \times \binom{N-s}{n-x}}{\binom{N}{n}}$$

where, N, the size of the population is 20

s, the number of errors in the population is 2

n, the sample size is 6

x, the number of errors in the sample is 0

We compute each combinatorial separately

$$\binom{s}{x} = \binom{2}{0} = \frac{2!}{0!(2-0)!} = \frac{2!}{2!} = 1$$

$$\binom{N-s}{n-x} = \binom{18}{6} = \frac{18!}{6!(18-6)!} = \frac{18!}{6!12!} = 18{,}564$$

$$\binom{N}{n} = \binom{20}{6} = \frac{20!}{6!(20-6)!} = \frac{20!}{6!14!} = 38{,}760$$

Thus, $P(X) = \dfrac{18564}{38760} = 0.4789$

Similarly, the probability of finding exactly one location with overstatement error in a sample of 6 can be computed as

$$\binom{s}{x} = \binom{2}{1} = \frac{2!}{1!(2-1)!} = \frac{2!}{1!} = 2$$

$$\binom{N-s}{n-x} = \binom{18}{5} = \frac{18!}{5!(18-5)!} = \frac{18!}{5!13!} = 8{,}568$$

$$\binom{N}{n} = \binom{20}{6} = \frac{20!}{6!(20-6)!} = \frac{20!}{6!14!} = 38{,}760$$

Thus, $P(x) = \dfrac{(2)(8568)}{38760} = 0.4422$

And the probability of finding both locations with overstatement error in the sample of 6 can be computed as

$$\binom{s}{x} = \binom{2}{2} = \frac{2!}{2!(2-2)!} = \frac{2!}{2!} = 1$$

$$\binom{N-s}{n-x} = \binom{18}{4} = \frac{18!}{4!(18-4)!} = \frac{18!}{4!14!} = 3{,}060$$

$$\frac{N}{n} = \frac{20}{6} = \frac{20!}{6!(20-6)!} = \frac{20!}{6!14!} = 38{,}760$$

Thus, $P(x) = \dfrac{3060}{38760} = 0.0789$

Note that x cannot be greater than 2 given there are only two inventory locations in the population with overstatement errors. Verify that the three probabilities computed here add up to one. That is, the three events are mutually exclusive and collectively exhaustive.

In an auditing and forensic accounting context, similar to this example, hypergeometric distribution may be the only distribution that is applicable. The probability concepts of hypergeometric distribution presented here and Bayesian probability revision presented in the previous section together could be very useful to auditors. In a forensic context, the accountant is less concerned with predicting the probability of finding no errors in the sample, but more interested in assessing the evidence of having found no errors after visiting six locations. The auditor has to assess the posterior probability that there are no material errors in locations he has not visited given the evidence that he did not find material errors in locations he did visit. The probabilities computed using this hypergeometric distribution provides the likelihoods that can then be used in Bayesian equation to revise the prior probabilities on the error rate at inventory locations.

7.6 Chapter Summary

This chapter presented the generic formula for computing the mean and standard deviation for any type of probability distribution, as well as introduced three widely applicable discrete probability distributions: the binomial, the Poisson, and the hypergeometric distributions. The next chapter discusses continuous probability distributions that build on the concepts presented in this chapter.

The three types of probability distributions presented in this chapter are of great relevance in auditing and forensic accounting. A forensic accountant has to sift through large amounts of financial data and is required to make probabilistic inferences. At times, it may not only be cumbersome but impossible to examine each data element. In such situations, a subset of the data is examined, such as in the case of multi-location inventory audits. When making inferences based on the properties of a specified probability distribution it is imperative to assess the underlying probability distribution correctly. Employing an incorrectly specified probability distribution would lead to fallacious inferences. Hence, for an accountant and auditor it is perhaps more important to know the applicability of various probability distributions than the computational mechanics associated with each. The computational mechanics are automated, and the appendix that follows provides a brief guide to using Excel to make the computations illustrated in this chapter.

In summary, when determining which of the three discrete probability distributions is most apt for the required task, the forensic accountant has to make some judgments related to the data-set. All of these distributions are used only in situations where the uncertainty can be represented as two choices: success or failure; error or no error; complies does not comply; occurs or does not occur. In situations when the complete count is available for only one of the outcomes, use of Poisson distribution is appropriate. The Poisson distribution is specified by only one parameter, λ. In the forensic accounting context, Poisson distribution is widely used in cyber-security to detect and prevent incidences of hacking. It is also applicable in continuous auditing situations. In situations where the population and sample are quite large and the probability of the outcomes is close to 0.5, the use of binomial distribution is appropriate. In such situations, the resultant binomial distribution is centered around the mean, and the mathematical properties for such distribution is greatly developed and widely available. However, in situations where the probability of outcomes are close to zero or one and the population is small, the underlying assumptions of binomial distributions are violated, hence its use will not be appropriate. In such situations, hypergeometric probability distribution is appropriate.

Appendix: Using Excel for Statistical Computations

Binomial Distribution:

For computations related to binomial distribution, use the BINOMDIST worksheet function. The result is the probability value of interest. The function is presented as BINOMIDIST (X, sample size, π, cumulative). X is the number of events of interest; that is, the specific value of the random variable X, in these illustrations presented as P(X= x). "Sample size" is self-evident. π is the probability of the event of interest. "Cumulative" is a "True" or "False" value. If interested in obtaining cumulative probability, $P(X \leq x)$, answer True. If interested in obtaining the probability value for only that event, $P(X = x)$, answer False.

Poisson Distribution:

Use the POISSON worksheet function to compute Poisson probabilities. Enter the function as POISSON(X, lambda, cumulative), where X is the number of events of interest. Lambda, λ, is the only parameter required to specify the Poisson distribution. Recall that λ is both the mean and variance of the Poisson distribution over unit time. "Cumulative" is a "True" or "False" value. If interested in obtaining cumulative probability, $P(X \leq x)$, answer True. If interested in obtaining the probability value for only that event, $P(X = x)$, answer False.

Hypergeometric Distribution:

Enter the function as HYPGEOMDIST (X, sample size, A, Population size). X is the number of events of interest; that is, the specific value of the random variable X, in our illustrations on multi-location audit the value of X was successively 0, 1, 2. "Sample size" is self evident and in our formulation it was denoted by "n." A is the number of events of interest in the population. In this formulation the term is represented by an "s," and in our illustrations of multilocation problem the value was 2. The "Population Size" is self-evident and was denoted as N in the formulation, and in the illustration the value was 20.

7.7 Endnotes

1. J. Cao, "Internet traffic tends toward Poisson and independent as the load increases," in *Non-Linear Estimation and Classification.*

2. A detailed description of the approach can be found at http://www2.ensc.sfu.ca/people/grad/pwangf/IPSW_report.pdf.

3. This problem is discussed at length in L. Graham, *Proceedings of the Symposium on the Extent of Testing in Auditing.*

8

Continuous Probability Distributions

Intuition is a poor guide when facing probabilistic evidence.

8.1 Introduction

The previous chapter presented concepts for discrete probability distributions, situations with a finite number of outcomes. In this chapter this discussion is extended to continuous probability distributions, or situations in which infinite outcomes are possible and the probability of any specific outcome is close to zero, hence of limited use. In an accounting and auditing context, the continuous probability distribution is relevant when the variables of interest are account balances. Account balances, or errors in those, presented in dollar amounts or percentages form, infinite continuous possibilities and are better represented through continuous probability distributions.

Continuous probability distributions are applicable in situations where the matter of interest is a range of values rather than a specific value. In an auditing context, the notion of materiality subsumes that a range of value is of greater applicability than a specific value. For example, the assertion that revenue of a company is fairly stated at $2 billion does not imply that the actual amount is *exactly* $2 billion, instead that it is within a narrow range of that value. The range of values is based on materiality judgment. In most situations, fair statement would imply that there is a high probability that the revenue is not mis-stated by more than $20 million, or within 1% of the stated amount. Conceptually, the cumulative probability of error less than $20 million is the sum of discrete probabilities of errors $0, $1, $2.... $20 million. Rather than estimating these individual probabilities and adding each discrete probability, it is more convenient to use continuous distribution and employ tools of calculus, such as integration. The preceding discussion is applicable when discussing errors in percentage terms. That is, it is of limited usefulness to precisely ascertain that the error in revenue is exactly 1%; instead, it is more useful to ascertain that error is not greater than 1%. In such situations the use of continuous probability eases the estimation and computation. Thus, in accounting and auditing context when making assertions related to account balances, continuous probabilities on ranges of value are more applicable than discrete probability on a specific value.

This chapter covers the concepts, mathematics, and applicability of continuous probability distributions. First is a discussion of these concepts in general, extending into commonly used continuous probability distributions: uniform distribution, normal distribution, and exponential distribution. The chapter concludes with joint probabilities of continuous variables.

8.2 Conceptual Development of Probability Framework

Two important mathematical concepts underlying continuous probability are probability distribution function, denoted as f(x), and the cumulative distribution function, denoted as F(x). The probability distribution function is a mathematical function of discrete probability of the variable of interest "x." Substituting the value of the variable "x = X" into the function enables computation of the discrete probability. As the distribution is considered to be continuous, the probability of each and every "x" contained in the relevant range is positive, and the probability for "x" outside the range is 0. The cumulative distribution function, on the other hand, denotes probability of the variable being less than or equal to a certain value X. That is, it estimates the probability that the variable of interest does not exceed X, or F(x) = P(x ≤ X).

The concepts here can be illustrated through a simple example of a gas station with a gas tank capacity of 1,000 gallons that gets replenished at the start of business each day. At the end of the day, the probability function that denotes the amount of gas sold during the day is a continuous function. Any specific amount, say 238.76 gallons, has negligible probability assigned to it and is of limited interest. It may be more pertinent to determine that the total sale was less than 250 gallons or between 300 and 400 gallons. These questions can be addressed through the knowledge of probability distribution function and cumulative distribution function. In this example, assume that there is no knowledge regarding the amount of sale on any particular day and that the sales of any amount between 0 gallons and 1,000 gallons are equally likely. That is, the probability of the sale of 500.00 gallons is the same as the sale of 824.98 gallons and is the same as that of 219.42 gallons. This is represented through a uniform probability distribution function with a probability distribution function of sale in gallons (rounded to the nearest whole number) as:

$f(x) = 0.001$ for $0 \leq x \leq 1{,}000$, else 0.

That is, the probability that the total sale is less than 0 or greater than 1,000 gallon is zero, but sale of less than 1,000 gallon has equal probability of 0.001. The sum of individual probabilities between 0 gallons and 1,000 gallons equal one, as these are mutually exclusive and collectively exhaustive. The probability distribution function yields a cumulative probability distribution of the sales for the day as follows:

$F(x) = 0$ if $x < 0$

$F(x) = 0.001x$ if $0 < x < 1{,}000$

$F(x) = 1$ if $x > 1{,}000$.

This implies that the probability of total sales being less than 0 gallons is 0 and the probability of the sales being less than or equal to 1,000 gallons, the capacity of the tank, is equal to one. The probability of sale of any amount less than "x" is 0.001x, that is the probability that the sale is less than 750 gallons is (0.001)(750) or 0.75. As expected, the function depicts that there is 50% probability that the sale is less than 500 gallons; (0.001)(500) equals 0.5. Cumulative distribution function enables the computation of probability that the sale is between 650 gallons and 800 gallons. Mathematically this is represented as

$P(650 \leq x \leq 800) = F(x \leq 800) - F(x \leq 650)$ or $0.8 - 0.65$ equals 0.15.

That is, the probability that sales are between 650 gallons and 800 gallons can alternatively be written as sales are less than 800 gallons but not less than 650 gallons. The cumulative probability that the sale is less than 800 gallons less the cumulative probability that it is less than 650 gallons yields the probability that the sale is between 650 gallons and 800 gallons. In this case, that probability is 0.15. In general,

$P(a \leq x \leq b) = F(b) - F(a)$

for any continuous probability distribution. The graphs of probability distribution function and cumulative distribution function for this example are shown in Figure 8.1.

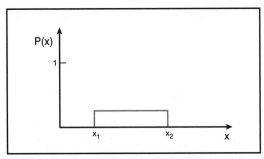

Panel A: Probability Distribution Function

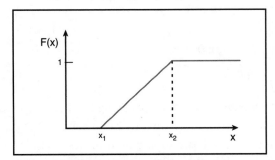

Panel B: Cumulative Distribution Function

Figure 8.1 Uniform distribution

Continuous probability distribution can be approximated from a discrete probability distribution on some discrete values. Extending the example of the gas station, suppose past sales have been between 400 gallons and 800 gallons a day with the highest occurrence being around 600 gallons. Further, the probability of sales increases linearly from 400 gallons to 600 gallons and then decreases linearly from 600 gallons to 800 gallons. That is, the probability distribution is in the shape of a triangle with 600 gallon being the apex. This probability distribution is shown in Panel A of Figure 8.2 and mathematically represented as:

$$f(x) = \frac{0.005}{200(x-400)} \text{ for } 400 \leq x \leq 600;$$

$$f(x) = 0.005 - \frac{0.005}{200}(x-600) \text{ for } 600 \leq x \leq 800;$$

$f(x) = 0$ for all other values of x.

The corresponding cumulative probability distribution is shown in Panel B of Figure 8.2.

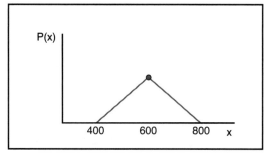

Panel A: Probability Distribution Function

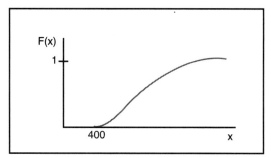

Panel B: Cumulative Distribution Function

Figure 8.2   Triangular probability distribution

The probability distribution function can be used to assess probability of various assertions, such as:

a. The probability of sales being greater than 680 gallons

b. The probability of sales being less than 560 gallons

c. The probability of sales being between 600 gallons and 700 gallons

d. The probability of sales being between 580 gallons and 640 gallons

These assertions are graphically shown in the four panels of Figure 8.3. The shaded area in each graph denotes the relevant probability.

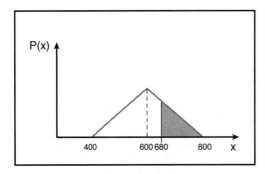

A. Probability of sales greater than 680 gallons is shaded region.

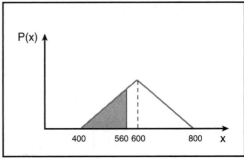

B. Probability of sales less than 560 gallons.

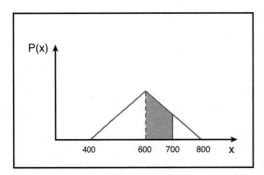

C. Probability of sales between 600 and 700 gallons.

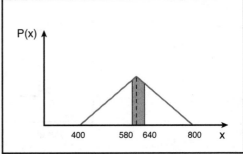

D. Probability of sales between 580 and 640 gallons.

Figure 8.3 Graphical Representation of Numerical Examples

The mathematical computation of the probabilities involves finding the area of relevant triangles. The area of the triangle is given by

Area of triangle = $0.5 \times (\text{base}) \times (\text{height})$

The solutions to these problems are as follows:

a. Probability of sales being greater than 680 gallons:

Base = $800 - 680 = 120$

$$\text{Height} = 0.005 - \frac{0.005}{200}(680 - 600) = 0.003$$

Hence, area of triangle, or cumulative probability = $\frac{1}{2}(120)(0.003) = 0.18$.

Thus, the probability that the sales are greater than 680 gallons is 0.18.

b. Probability of sales being less than 560 gallons is

Base = 560 − 400 = 160

$$\text{Height} = \frac{0.005}{200}(560-400) = 0.004$$

Hence, area of triangle, or cumulative probability = $\frac{1}{2}(160)(0.004) = 0.32$.

Thus, the probability that the sales are less than 560 gallons is 0.32.

c. Probability of sales being between 600 gallons and 700 gallons.

$P(600 \le x \le 700) = 1 - P(x \le 600) - P(x > 700)$.

$P(x \le 600)$ or probability that sales are less than 600 gallons can be computed as

Base = 600 − 400 = 200

$$\text{Height} = \frac{0.005}{200}(600-400) = 0.005$$

Hence, area of triangle, or cumulative probability = $\frac{1}{2}(200)(0.005) = 0.5$.

Thus, the probability that the sales are less than 600 gallon is 0.5.

$P(700 \le x)$ or probability that sales are greater than 700 gallons can be computed as

Base = 800 − 700 = 100

$$\text{Height} = 0.005 - \frac{0.005}{200}(700-600) = 0.0025$$

Hence, area of triangle, or cumulative probability = $\frac{1}{2}(100)(0.0025) = 0.125$.

Thus, the probability that the sales are greater than 700 gallon is 0.125.

$P(600 \le x \le 700) = 1 - P(x \le 600) - P(700 \le x)$, or $1 - 0.5 - 0.125 = 0.375$

Thus, the probability that sales are between 600 gallons and 700 gallons is 0.375.

d. Probability of sales being between 580 gallons and 640 gallons.

$P(580 \leq x \leq 640) = 1 - P(x \leq 580) - P(640 \leq x)$.

$P(x \leq 580)$ or probability that sales are less than 580 gallons can be computed as

$$\text{Height} = \frac{0.005}{200}(580 - 400) = 0.0045$$

Hence, area of triangle, or cumulative probability = $^1/_2$ (180) (0.0045) = 0.405.

Thus, the probability that the sales are less than 580 gallons is 0.405.

$P(640 \leq x)$ or probability that sales are greater than 640 gallons can be computed as

Base = 800 − 640 = 160

$$\text{Height} = 0.005 - \frac{0.005}{200}(640 - 600) = 0.004$$

Hence, area of triangle, or cumulative probability = $^1/_2$ (160) (0.004) = 0.32.

Thus, the probability that the sales are greater than 640 gallons is 0.32.

$P(580 \leq x \leq 640) = 1 - P(x \leq 580) - P(640 \leq x)$, or 1 − 0.405 − 0.32 = 0.275

Thus, the probability that sales are between 580 gallons and 640 gallons is 0.275.

8.3 Uniform Probability Distribution

In the previous section you learned about uniform probability distribution in the numerical example. Next is the mathematical development of the concept. Uniform distribution is sometimes referred to as the rectangular distribution because each value within a range has equal probability of occurrence. The probability of occurrence outside the range is zero. The probability within the range equals the reciprocal of the width of the range. Thus, the probability distribution function of a uniform distribution over the range {a, b} is given as

$$f(x) = \frac{1}{(b-a)} \text{ for } a \leq x \leq b$$

$f(x) = 0$ for all other values of x

The probability distribution function can be used to find probability of specific assertions. For example, the mean of the distribution is at the mid-point of the range or at $\frac{a+b}{2}$, which is also the median. Thus, uniform distribution is symmetrical, as the mean equals the median. The variance of a uniform distribution is

$$\sigma_x^2 = \Sigma[(x - \mu_x)^2] = \frac{(b-a)^2}{12}$$

and the standard deviation is

$$\sigma_x = \sqrt{\frac{(b-a)^2}{12}}$$

The uniform distribution is represented as a rectangle, and computation of cumulative probability is relatively simple. Being a rectangular distribution the height, which denotes the probability, is constant. The cumulative probability or the area of the rectangle is determined solely by the length of the base. As the entire rectangle has an area of one, the area of any portion can be computed by taking the proportion of the base of the specified area to the range, (b-a). That is, the probability that the event lies between "c" and "d", where $a \leq c \leq d \leq b$, equals $\frac{(d-c)}{(b-a)}$. For cumulative probability computations, "c" is replaced by "a" in the equation, yielding $\frac{(d-a)}{(b-a)}$.

8.4 Normal Probability Distribution

One of the most commonly observed random variables has a bell-shaped probability distribution as shown in Figure 8.4. The probability distribution is commonly known as the normal probability distribution or the Gaussian probability distribution named after the famous statistician Gauss. This distribution plays a critical role in the science of inference.

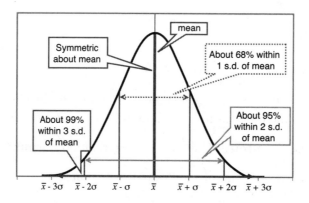

Figure 8.4　Plot of a normal distribution

Many business events generate random variables with probability distributions that are well approximated by a normal distribution. For example, the patterns of stock and bond prices are often modeled using normal distribution. Sales and production also follow a normal distribution pattern. Additionally, macro-economic models use normal distribution. These lead to wide-ranging applications of normal distribution in business and accounting context.

The applicability and importance of normal distribution increases because it can be used for sampling even in distributions which are known to be not normal. The *Central Limit Theorem* proves that as the sample size gets large enough, the sample mean is normally distributed, regardless of the underlying distribution of the population. That is, if one makes multiple sets of 40 observations from a non-normal distribution, the mean of each set of 40 observations will be normally distributed. This theorem and its critical implications are further discussed in Chapter 11, "Determining Sample Size."

The normal distribution has several important theoretical properties that enhance its applicability and ease of mathematical computation. First, it is a symmetrical distribution, thus the mean of the distribution equals the median. Second, it is bell-shaped in appearance, thus observations closer to the mean are more likely than those further from it. In fact, 50% of the values in normal distribution are contained within two-thirds of standard deviation around the mean. Third, it has an infinite range though the probability at the extreme tails will be negligible due to the bell-shaped nature of the curve. However non-zero probability at the tails allows the extreme observations to be plausible.

For normal distribution, the mathematical expression for probability density function is given as

$$f(X) = \frac{1}{\sqrt{2\pi}\sigma} e^{-\frac{1}{2}\left[\frac{X-\mu}{\sigma}\right]^2}$$

where, e = the mathematical constant approximated by 2.7183

π = the mathematical constant approximated by 3.1416

μ = the mean of the normal distribution

σ = the standard deviation of the normal distribution

X = any value of the continuous variable, where $-\infty < X < \infty$.

In the above formula, e and π are mathematical constants, hence the probabilities of X is dependent on two parameters, the mean μ and the standard deviation σ. For any distribution, once the value of μ and σ are specified the distribution is uniquely generated. Mathematically, a random variable is shown to be normally distributed with a mean of μ and a variance of σ^2 as

$$X \sim N(\mu, \sigma^2)$$

Consider the previous example regarding gas sales from a tank with the maximum capacity of 1,000 gallons. Let's assume that the amount of sale is normally distributed. Figure 8.5 illustrates the normal distribution for four different assumptions of mean and standard deviation. Plot A shows the normal distribution for the baseline case, assuming a mean of 600 gallons and the standard deviation of 100 gallons. Plot B shows the normal distribution keeping the mean at 600 gallons but increasing the standard deviation to 150 gallons. As can be seen, Plot A and B are centered around the same mean, but Plot B is wider and flatter due to the higher standard deviation. Plot C keeps the standard deviation at 100 gallons but assumes a mean of 400 gallons. Comparing Plot A and C, it can be seen that the shapes are identical, except Plot C shifts to the left due to the lower mean. Plot D shows the normal distribution with the mean of 600 gallons and a standard deviation of 50 gallons. Comparing Plot D to Plot A, it can be seen that Plot D is narrower and taller to signify a smaller standard deviation.

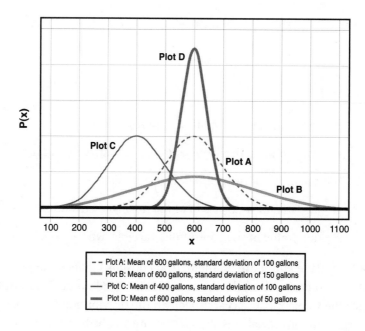

Figure 8.5 Normal distributions with varying means and standard deviation

As discussed earlier, the probability of any specific value of the random variable X is negligible in a continuous probability distribution. Hence, probability assertions are made for a range of values of X, that is $x_1 \leq X \leq x_2$. Mathematically, this is computed by integrating this probability distribution function with a lower limit of x_1 and an upper limit of x_2. That is,

$$P(x_1 \leq X \leq x_2) = \int_{x_1}^{x_2} \frac{1}{\sqrt{2\pi}\sigma} e^{-\frac{1}{2}\left[\frac{X-\mu}{\sigma}\right]^2} dx$$

Expressing this in terms of cumulative distribution function,

$$P(x_1 \leq X \leq x_2) = P(X \leq x_2) - P(X \leq x_1),$$

or

$$P(x_1 \leq X \leq x_2) = \int_0^{x_2} \frac{1}{\sqrt{2\pi}\sigma} e^{-\frac{1}{2}\left[\frac{X-\mu}{\sigma}\right]^2} dx - \int_0^{x_1} \frac{1}{\sqrt{2\pi}\sigma} e^{-\frac{1}{2}\left[\frac{X-\mu}{\sigma}\right]^2} dx$$

Though the mathematical expressions here require solving integral calculus to compute probabilities, widely available normal probability tables are easier to use in practice. Further, software such as Microsoft Excel have these computations automated as discussed in the Appendix. The cumulative probability distribution tables for normal distribution are for N(0, 1), that is a normal distribution with a mean of 0 and standard deviation of 1, also known as the standardized normal distribution. All other normal distributions are converted to the standard normal through a transformation process. Any normally distributed random variable X is first converted to a standardized normal random variable Z. The Z value is equal to the difference between X and the mean µ, divided by the standard deviation, σ. That is,

$$Z = \frac{(X-\mu)}{\sigma}$$

Although the random variable X is from distribution N(µ, σ²), that is a normal distribution with mean µ and standard deviation σ, the standardized variable Z has a mean of 0 and standard deviation of 1. After the variable X is converted into its standardized form, the probabilities can be determined from Table 8.1, the *cumulative standardized normal distribution*.

The transformational formula is demonstrated through this gas tank example. Suppose the sales are normally distributed with a mean of 600 gallons and a standard deviation of 75 gallons, that is, the probability distribution is N (600, 5625). Compute the following:

a. The probability that the sales are less than 560 gallons.

b. The probability that the sales are greater than 720 gallons.

c. The probability that the sales are between 580 gallons and 700 gallons.

d. The probability that the sales are between 520 gallons and 560 gallons.

Table 8.1 Standard Normal Distribution Table

z	0	0.01	0.02	0.03	0.04	0.05	0.06	0.07	0.08	0.09
0	0.5	0.504	0.508	0.512	0.516	0.5199	0.5239	0.5279	0.5319	0.5359
0.1	0.5398	0.5438	0.5478	0.5517	0.5557	0.5596	0.5636	0.5675	0.5714	0.5753
0.2	0.5793	0.5832	0.5871	0.591	0.5948	0.5987	0.6026	0.6064	0.6103	0.6141
0.3	0.6179	0.6217	0.6255	0.6293	0.6331	0.6368	0.6406	0.6443	0.648	0.6517
0.4	0.6554	0.6591	0.6628	0.6664	0.67	0.6736	0.6772	0.6808	0.6844	0.6879
0.5	0.6915	0.695	0.6985	0.7019	0.7054	0.7088	0.7123	0.7157	0.719	0.7224
0.6	0.7257	0.7291	0.7324	0.7357	0.7389	0.7422	0.7454	0.7486	0.7517	0.7549
0.7	0.758	0.7611	0.7642	0.7673	0.7704	0.7734	0.7764	0.7794	0.7823	0.7852
0.8	0.7881	0.791	0.7939	0.7967	0.7995	0.8023	0.8051	0.8078	0.8106	0.8133
0.9	0.8159	0.8186	0.8212	0.8238	0.8264	0.8289	0.8315	0.834	0.8365	0.8389
1.0	0.8413	0.8438	0.8461	0.8485	0.8508	0.8531	0.8554	0.8577	0.8599	0.8621
1.1	0.8643	0.8665	0.8686	0.8708	0.8729	0.8749	0.877	0.879	0.881	0.883
1.2	0.8849	0.8869	0.8888	0.8907	0.8925	0.8944	0.8962	0.898	0.8997	0.9015
1.3	0.9032	0.9049	0.9066	0.9082	0.9099	0.9115	0.9131	0.9147	0.9162	0.9177
1.4	0.9192	0.9207	0.9222	0.9236	0.9251	0.9265	0.9279	0.9292	0.9306	0.9319
1.5	0.9332	0.9345	0.9357	0.937	0.9382	0.9394	0.9406	0.9418	0.9429	0.9441
1.6	0.9452	0.9463	0.9474	0.9484	0.9495	0.9505	0.9515	0.9525	0.9535	0.9545
1.7	0.9554	0.9564	0.9573	0.9582	0.9591	0.9599	0.9608	0.9616	0.9625	0.9633
1.8	0.9641	0.9649	0.9656	0.9664	0.9671	0.9678	0.9686	0.9693	0.9699	0.9706
1.9	0.9713	0.9719	0.9726	0.9732	0.9738	0.9744	0.975	0.9756	0.9761	0.9767
2.0	0.9772	0.9778	0.9783	0.9788	0.9793	0.9798	0.9803	0.9808	0.9812	0.9817
2.1	0.9821	0.9826	0.983	0.9834	0.9838	0.9842	0.9846	0.985	0.9854	0.9857
2.2	0.9861	0.9864	0.9868	0.9871	0.9875	0.9878	0.9881	0.9884	0.9887	0.989
2.3	0.9893	0.9896	0.9898	0.9901	0.9904	0.9906	0.9909	0.9911	0.9913	0.9916
2.4	0.9918	0.992	0.9922	0.9925	0.9927	0.9929	0.9931	0.9932	0.9934	0.9936
2.5	0.9938	0.994	0.9941	0.9943	0.9945	0.9946	0.9948	0.9949	0.9951	0.9952
2.6	0.9953	0.9955	0.9956	0.9957	0.9959	0.996	0.9961	0.9962	0.9963	0.9964
2.7	0.9965	0.9966	0.9967	0.9968	0.9969	0.997	0.9971	0.9972	0.9973	0.9974
2.8	0.9974	0.9975	0.9976	0.9977	0.9977	0.9978	0.9979	0.9979	0.998	0.9981
2.9	0.9981	0.9982	0.9982	0.9983	0.9984	0.9984	0.9985	0.9985	0.9986	0.9986
3.0	0.9987	0.9987	0.9987	0.9988	0.9988	0.9989	0.9989	0.9989	0.999	0.999

The normal probability distribution N (600, 5625) with the corresponding region shaded are illustrated in Panels A–D of Figure 8.6. The steps to solve these problems require identifying the relevant region for which cumulative probabilities are to be determined; transforming the random variable to equivalent standardized normal or z-value; finding the cumulative area under the normal curve (use Table 8.1); arithmetic manipulation (addition or subtraction) to determine the probability of the shaded region. Compute the probabilities as the following:

a. The probability that the sales are less than 560 gallons.

The area of interest is shaded in Panel A of Figure 8.6.

$X = 560$, $\mu = 600$ and $\sigma = \sqrt{5625} = 75$

Thus, $z = \dfrac{(560-600)}{75} = -0.5333$

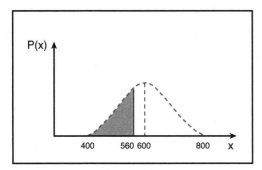

A. Probability of sales less than 560 gallons.

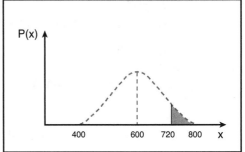

B. Probability of sales greater than 720 gallons.

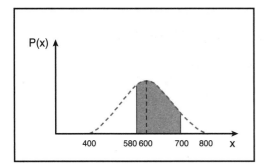

C. Probability of sales between 580 and 720 gallons.

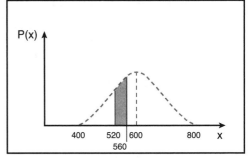

D. Probability of sales between 520 and 560 gallons.

Figure 8.6 Graphical representation of numerical examples using normal distribution

To use Table 8.1 to find the cumulative area under the normal curve, z-value of interest has to be located to the nearest 1/10. For positive values of z, the cumulative probability can be obtained directly from the table. For negative values of z, the cumulative probability is obtained by subtracting the number from the table corresponding to the absolute value of z from 1. You can scan down the Z column until you get to the row corresponding to 0.5; then go across the columns to 0.03. The body of the table gives the corresponding cumulative probability, which in this case is 0.7019. Subtracting that from 1, we get 0.2981.

Because the question of interest was to determine the probability that the sales are less than 560 gallons, the answer is 0.2981.

b. The probability that the sales are greater than 720 gallons.

The area of interest is shaded in Panel B of Figure 8.6.

$X = 720$, $\mu = 600$ and $\sigma = \sqrt{5625} = 75$

Thus, $z = \dfrac{(720-600)}{75} = 1.6000$

To use Table 8.1 to find the cumulative area under the normal curve, z-value of interest has to be located to the nearest 1/10. You can scan down the Z column until you get to the row corresponding to 1.6; then go across the columns to 0.00. The body of the table gives the corresponding cumulative probability, which in this case is 0.9452.

Because the question of interest was to determine the probability that the sales are greater than 720 gallons, the answer is 1 − 0.9452 or 0.0548.

c. The probability that the sales are between 580 gallons and 700 gallons.

The area of interest is shaded in Panel C of Figure 8.6.

$X_1 = 580$, $X_2 = 700$ $\mu = 600$ and $\sigma = \sqrt{5625} = 75$

Thus, $z_1 = \dfrac{(580-600)}{75} = -0.2666$

and $z_2 = \dfrac{(700-600)}{75} = 1.333$

To use Table 8.1 to find the cumulative area under the normal curve, the cumulative probability corresponding to the z-value of −0.2666 is 0.3974 (or 1− 0.6064), and that corresponding to 1.333 is 0.9082.

Since the question of interest was to determine the probability that the sales are between 580 gallons and 700 gallons, it can be mathematically expressed as

$$P(580 \leq X \leq 700) = P(X \leq 700) - P(X \leq 580)$$

or

$$0.9082 - 0.3974 = 0.5108$$

Thus, the probability that the sales are between 580 gallons and 700 gallons is 0.5108.

d. The probability that the sales are between 520 gallons and 560 gallons.

The area of interest is shaded in Panel D of Figure 8.6.

$X_1 = 520$, $X_2 = 560$ $\mu = 600$ and $\sigma = \sqrt{5625} = 75$

Thus, $z_1 = \dfrac{(520 - 600)}{75} = -1.0666$

and $z_2 = \dfrac{(560 - 600)}{75} = -0.5333$

To use Table 8.1 to find the cumulative area under the normal curve, the cumulative probability corresponding to the z-value of −1.0666 is 0.1446, and that corresponding to −0.5333 is 0.2981.

Because the question of interest was to determine the probability that the sales are between 520 gallons and 560 gallons, it can be mathematically expressed as

$$P(520 \leq X \leq 560) = P(X \leq 560) - P(X \leq 520)$$

or

$$0.2981 - 0.1446 = 0.1535$$

Thus, the probability that the sales are between 520 gallons and 560 gallons is 0.1535.

These numerical examples illustrate how the value of a normal distribution clusters around the mean. For any normal distribution,

- 68.26% of the values fall within 1 standard deviation of the mean. That is, the area under the bell curve covering the range $(\mu - \sigma)$ to $(\mu + \sigma)$ is 0.6826.

- 95.44% of the values fall within 2 standard deviations of the mean. That is, the area under the bell curve covering the range ($\mu - 2\sigma$) to ($\mu + 2\sigma$) is 0.9544.
- 99.73% of the values fall within 3 standard deviations of the mean. That is, the area under the bell curve covering the range ($\mu - 3\sigma$) to ($\mu + 3\sigma$) is 0.9973.

Another type of probability problems involves determining the value of X associated with a known probability. In this example, these problems can be constructed as follows:

a. There is a 10% probability that the sales will be less than X gallons.

b. There is a 25% probability that the sales will be greater than X gallons.

c. There is a 90% probability that the sales will be between 600 ± X gallons.

The steps required to find a particular value associated with a known probability are to sketch the normal curve and place the values for the mean and X; next find the cumulative area less than X; shade the area of interest; using the table determine the corresponding area under the normal curve less than X; and finally use the equation here to solve for X:

$$X = \mu + Z\sigma$$

The normal probability plots corresponding to these three questions are drawn in Panels A, B, and C of Figure 8.7. The solutions are

a. There is a 10% probability that the sales will be less than X gallons.

$\mu = 600$, $\sigma = \sqrt{5625}$ or 75, probability is 0.1

The z-value corresponding to a cumulative probability of 0.1 can be located from Table 8.1. You locate a number closest to 0.100 in the body of the table, and the corresponding row and column determine Z. Since the table starts with 0.500, the probability of 0.1 is not available in Table 8.1. Instead, we find the z-value corresponding to 0.9 (1–0.1) and multiply it with –1 to get the corresponding z-value. From Table 8.1, the z-value closest to 0.9 is 0.8997, which is in the row for 1.2 and column for 0.08, yielding a z-value of 1.28. We multiply 1.28 with –1, and the z-value corresponding to 0.1 is –1.28.

Substituting the Z-value just given along with the mean and standard deviation for the problem, you can solve for X:

X = 600 – 1.28 (75) = 504 gallons.

Or there is a probability of 0.1 that the sales would be less than 504 gallons.

b. There is 25% probability that the sales will be greater than X gallons.

$\mu = 600$, $\sigma = \sqrt{5625}$ or 75, probability of sales being greater than X is 0.25.

If the probability of sales being greater than X is 0.25, then the probability of it being less than X is 0.75. Or the cumulative probability of X is 0.75.

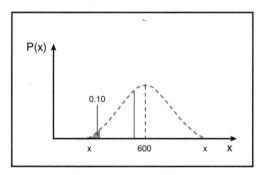
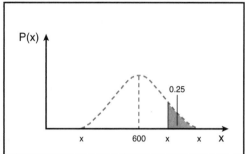

A. 10% probability sales will be less than X gallons.

B. 25% probability sales will be greater than X gallons.

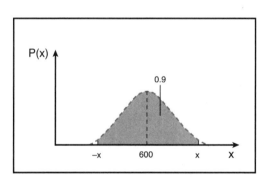

C. 90% probability sales will be between 600 − x and 600 + x gallons.

Figure 8.7 Graphical representation of numerical examples using normal distribution

The z-value corresponding to a cumulative probability of 0.75 can be located from Table 8.1. You can locate 0.7486 corresponding to row 0.6 and column 0.07. The next value is 0.7518 corresponding to the column 0.08. From this, we can approximate that the z-value corresponding to 0.75 is 0.675.

Substituting the above Z-value along with the mean and standard deviation for the problem, you can solve for X:

X = 600 + 0.675 (75) = 650.625 gallons

Or there is a probability of 0.25 that the sales would be greater than 650.625 gallons.

c. There is 90% probability that the sales will be between 600 ± X gallons.

$\mu = 600$, $\sigma = \sqrt{5625}$ or 75, and probability is 90%

Because the distribution is symmetric, the two tails have equal probability of 0.05. The z-value corresponding to a cumulative probability of 0.95 can be located from Table 8.1. You can locate a number closest to 0.95 in the body of the table, and the corresponding row and column determine Z. For example, 0.9495 is located at the row with Z-value of 1.6, and the column 0.04 and 0.9505 is under the column 0.05, yielding a z-value of −1.645.

Substituting this Z-value along with the mean and standard deviation for the problem, you can solve for X:

X = 600 ± 1.645 (75) = 476.635 and 723.375.

Or there is a probability of 0.9 that the sales are between 476.635 gallons and 723.375 gallons.

8.5 Testing for Normality

Prior to using techniques and tables of normal distribution, it is important to ascertain whether or not the underlying characteristics and assumptions of normal distribution are met by the data-set. For situations involving actual accounting data, you would like to know if the data has come from a distribution that approximates a normal distribution to ensure valid application of normality principles. There are two ways to determine whether a set of data is normally distributed. First is through a comparison of data characteristics to the theoretical properties of a normal distribution. Second is through constructing a normal probability plot.

As discussed in a previous section, normal distribution has important theoretical characteristics. Because the distribution is symmetric, the mean equals the median. It is relatively easy to evaluate whether or not a data-set meets that criteria. If the mean and median are very different, the distribution is not symmetric, thereby limiting the applicability of normal distribution to that context. Second, approximately 68% of the observation is expected to fall within one standard deviation of the mean. This follows from the inter-quartile property of the bell curve. Third, there should be extremely few observations that are three standard deviations from the mean. In other words, the range of the data-set should not exceed six standard deviations. Computing descriptive statistics from the data-set and comparing those statistics to the theoretical properties of the normal distribution helps ascertain the appropriateness of using normal distribution to approximate the data-set.

Additional distributional properties of normality can be tested for the data-set. That is, the standard deviation and the interquartile range could be computed for the data-set. If the data-set is normally distributed, the interquartile range should be close to 1.33 standard deviation. The departure from the theoretical relationship provides a basis to assess whether or not normal distribution is a good approximation of the data-set. Similarly, the numbers that fall within one standard deviation from the mean can be assessed. Rarely would any data-set comply with all the theoretical properties of normal distribution. However, when the deviations from theoretical properties are minor, normal distribution could provide a good approximation.

Normal probability plot enables a visual evaluation of whether the data are normally each ordered value. For example, if there are 24 observations, there are 24 z-values computed corresponding to cumulative probabilities of 1/25, 2/25, 3/25,... 23/25, 24/25. These z-values are plotted on the x-axis, and the actual values of the variable are plotted on the y-axis. If the resultant plot approximates a straight line, the data are normally distributed. In situations when the plot is concave down (as in Panel A of Figure 8.8), it is left-skewed or skewed toward the lower values. Concave down implies that the data rises fast initially and then tapers off. In situations when the plot is concave upward, as in Panel C of Figure 8.8, the data is right-skewed or has few extremely large values. Concave up implies that the data rises slowly at the beginning and then rises at a faster rate as the higher values get plotted.

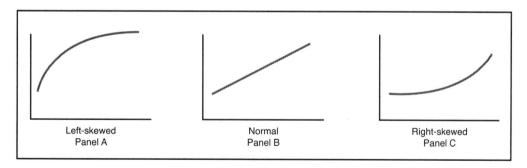

Figure 8.8 Testing for normality

Levine, David M.; Stephan, David F.; Krehbiel, Timothy C.; Berenson, Mark L.; *Statistics for Managers Using Microsoft Excel*, 7th Ed., ©2014, p. 235. Reprinted and electronically reproduced by permission of Pearson Education, Inc., Upper Saddle River, New Jersey.

8.6 Chebycheff's Inequality

In situations when the distribution is unknown or is non-normal, Chebycheff's inequality provides a conservative measure of cumulative probability. Chebycheff's inequality determines the minimum area that is enclosed around the mean. The inequality states

$$\text{Area} \geq 1 - \frac{1}{K^2}$$

where K is the number of standard deviations. Using this relationship a comparison can be made with the area under the normal distribution. As the method used in Chebycheff's inequality is based only on the measure of standard deviation, the inequality holds true for any probability distribution. In comparing the areas under the normal curve with that predicted by Chebycheff's inequality in Table 8.2, note that Chebycheff's inequality is the conservative measure and hence would have the lower value.

Table 8.2 Comparison of Normal Distribution with Chebycheff Inequality

Number of Standard Deviations	Normal Distribution	Chebycheff's Inequality
1	68.26%	-
1.5	86.64%	55.56%
2	95.44%	75.0%
2.5	98.76%	84.0%
3	99.74%	88.9%
4	99.9%	93.75%
5	100%	96.0%

The conservative estimate provided by Chebycheff's inequality is useful in an accounting and auditing context in which distributions are sometimes unknown. Consider the situation when an auditor has sampled a population and has determined the error rate in the sample, but his task is to estimate the maximum error rate in the population. Assuming normality, the auditor would be 99.7% certain that the error rate will not exceed three standard deviations from the mean. However, a more conservative estimate would be 4 standard deviations, and the theoretical underpinning is provided by the Chebycheff's inequality that the probability of error being greater than 4 standard deviations is 6.25%. This estimate, though more conservative than that of normal distribution, requires no assumptions regarding underlying probability distribution of the error rate.

8.7 Binomial Distribution Expressed as a Normal Distribution

Normal distribution approximates the pattern of discrete binomial distribution for repeated trials. That is, the number of heads in repeated 100 coin tosses follows a normal distribution. This approximation can be used to compute probabilities for large sample sizes. This approximation of applying techniques of normal distribution to binomial trials has wide applicability in forensic accounting and auditing to assess error rates, misstatements, and irregularities.

Let's consider an inventory audit, where the auditor ascertains whether an item of inventory exists or doesn't exist. The binomial random variable X is expressed as the sum of independent observations of inventory items:

$$X = X_1 + X_2 + \ldots + X_n$$

where X_i takes the value 1 if an error is detected and a value of 0 otherwise. The value of X is an estimate of the error rate in inventory items. If the probability of error is "p", then the probability of no error is "1− p". The number of errors found in the sample, X, has a mean and variance of

Mean = μ = np,

Variance = σ^2 = np(1− p).

The plot of a binomial distribution with p equal to 0.5 and n equal to 100 is shown in Figure 8.9. The plot has the same bell-shape as a normal distribution. This visual evidence is corroborated by statistical techniques outlined in the previous section to test for normality of repeated binomial trials. Further, as is discussed in a later section, the Central Limit Theorem provides the theoretical basis for approximating binomial distribution with normal for large numbers of repeated trials. Although the rule of thumb is that a sample size of 30 for a binomial distribution validates normality assumption, it is true when the probability p is between 0.2 and 0.8. For smaller probability values, as in the case with most forensic accounting problems, a higher sample size is required to validate normality assumption.

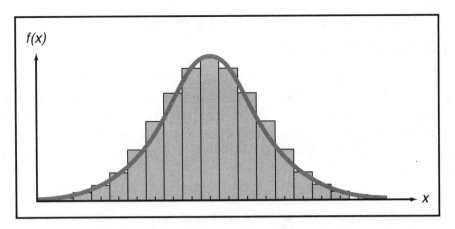

Binomial with p=0.5 and n=100 superimposed by a normal with μ=50 and σ=5

Figure 8.9 Normal as an approximation of a binomial distribution

McClave, James T.; Benson, P. George; Sincich, Terry L.; *A First Course in Business Statistics*, 8th Ed., ©2001, p. 213. Reprinted and electronically reproduced by permission of Pearson Education, Inc., Upper Saddle River, New Jersey.

8.8 The Exponential Distribution

Another commonly used continuous probability distribution is the exponential distribution. Exponential distribution is often used to model data that is right-skewed, or there is small probability of very large values. This distribution is commonly used in modeling waiting times and can sometimes be used in modeling account balances when there are numerous small balances and very few large account balances. The exponential distribution is defined by a single parameter, its mean, represented as λ. The probability distribution function is given as

$f(X) = \lambda e^{-\lambda x}$ for X > 0, where

X is the random variable 0 < X < ∞

λ is the mean number of arrivals per unit of time

and

e = 2.7183

The cumulative distribution function is as follows:

$f(X) = 1 - e^{-\lambda x}$, for x > 0.

This distribution has a mean equal to λ and a variance of $1/\lambda^2$ and is widely applicable in queuing, often used to represent the length of time between arrivals or to represent service time. The model assumptions are similar to that of Poisson distribution discussed in the previous chapter on discrete variables. This distribution provides the probability that a success, error, or arrival will occur during an interval of time x. Figure 8.10 plots the exponential distribution function for λ equal to 0.25. The area to the left of 8 on the x-axis gives the cumulative probability that the arrival will occur before 8 minutes. As before, the probability that the arrival is between 8 and 10 minutes can be computed:

P (8 ≤ x ≤ 10) = F (x < 10) − F (x < 8)

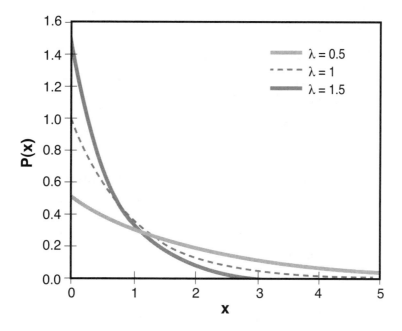

Probability distribution function of an exponential distribution with λ equal to 0.5, 1, and 1.5.

Figure 8.10   Plot of exponential distribution

8.9   Joint Distribution of Continuous Random Variables

In Chapter 6, "Transitioning to Evidence," the concept of joint probability was developed and illustrated. Similarly, many continuous variables can be modeled using joint probabilities. There are many applications in finance such as determining returns of the portfolio where the return of each stock can be viewed as random variables. Similarly in accounting and auditing various account balances or errors thereof are modeled as random variables and the auditor is interested in the joint probability distribution.

Let $X_1, X_2, ..., X_n$ be continuous random variables, say errors in account balances 1, 2,...n. The joint cumulative distribution function $F(x_1, x_2, ..., x_n)$ defines the probability that the error in each account balance is simultaneously less than x_1, x_2, etc. That is, error in account 1 is less than x_1, error in account 2 is less than x_2, and so on. Mathematically,

$$F(x_1, x_2, ..., x_n) = P(X_1 < x_1 \cap X_2 < x_2 \cap ... \cap X_n < x_n)$$

The individual random variables each have a cumulative distribution function also called their marginal distribution. That is, $F(x_1)$ indicates that the cumulative probability of error in Account 1 does not exceed x_1. In general,

$$F(x_i) = P(X_i < x_i) \text{ for all i's.}$$

The joint cumulative probability is a product of individual cumulative probabilities if the random variables are independent. That is,

$$F(x_1, x_2, ..., x_n) = F(x_1) F(x_2) ... F(x_n)$$

When the errors in account balances 1, 2, ... n are independent.

In general, however, the random variables need not be independent. For example, price changes of individual stocks are usually not independent. Instead when the market index is on an uptick, prices of stocks with positive beta are expected to increase. This dependence is modeled in statistics through the concept of covariance and correlation. These concepts are discussed in a later chapter; for the time being, we limit the discussion to independent random variables.

In some situations, the sum or differences of random variables may be of interest. That is, instead of ascertaining that the errors in individual accounts are less than a respective amount, you might be interested in ascertaining the sum of errors in two or more random variables. Consider $X_1, X_2, ..., X_n$ are continuous random variables, with means $\mu_1, \mu_2, ... \mu_n$ and variances $\sigma_1^2, \sigma_2^2, ... \sigma_n^2$. The mean of the sum of these n random variables is the sum of the individual means for each variable. Mathematically,

$$E(X_1 + X_2 + ... + X_n) = \mu_1 + \mu_2 + ... + \mu_n$$

and assuming that the variables are independent, the variance of the sum is the sum of the individual variances, given as

$$\text{Var}(X_1 + X_2 + ... + X_n) = \sigma_1^2 + \sigma_2^2 + ... + \sigma_n^2$$

Consequently, the standard deviation of the sum is the square root of the sum of squared individual standard deviations $\sigma_1, \sigma_2, ... \sigma_n$. That is,

$$\text{Std. Dev.}(X_1 + X_2 + ... + X_n) = \sqrt{\sigma_1^2 + \sigma_2^2 + ... + \sigma_n^2}$$

Similarly, the difference between a pair of random variables with means μ_1 and μ_2 and variances σ_1^2 and σ_2^2 can be computed as follows:

$$E(X_1 - X_2) = \mu_1 - \mu_2,$$

and assuming that the variables are independent, the variance of the sum is the sum of the individual variances, given as

$$\text{Var}(X_1 - X_2) = \sigma_1^2 + \sigma_2^2.$$

Consequently, the standard deviation of the sum is the square root of the sum of squared individual standard deviations $\sigma_1, \sigma_2, \ldots \sigma_n$. That is,

$$\text{Std. Dev.}(X_1 - X_2) = \sqrt{\sigma_1^2 + \sigma_2^2}$$

Now for a numerical example to illustrate this concept of joint continuous probability distribution. Suppose a contractor is suspicious of his supervisor at a site who has billed $360,000 for material and labor costs on a project. The contractor believes that the material cost is normally distributed with a mean of $140,000 and a standard deviation of $12,000; similarly the labor cost is normally distributed with a mean of $160,000 and a standard deviation of $15,000. Further, the two costs are independent. Recall, a normal distribution is represented as $N(\mu, \sigma^2)$. Mathematically, the probability distribution of material costs, M is $N(140{,}000, 144{,}000{,}000)$ and that of labor costs, L is $N(160{,}000, 225{,}000{,}000)$. Aggregate the two probability distributions to find the probability distribution for the total costs, T:

$$\mu_T = E(M + L) = \mu_M + \mu_L = 140{,}000 + 160{,}000 = \$300{,}000$$

and assuming that the material costs and labor costs are independent, the variance of the sum is the sum of the individual variances, given as

$$\sigma_T^2 = \text{Var}(X_1 + X_2) = \sigma_M^2 + \sigma_L^2 = 144{,}000{,}000 + 225{,}000{,}000 = 369{,}000{,}000$$

and the standard deviation of the total cost will be

$$\sqrt{369{,}000{,}000} = \$19{,}209$$

Alternatively, std. dev $\sqrt{(12{,}000)^2 + (15{,}000)^2} = \sqrt{369{,}000{,}000} = 19{,}209$

As the material costs and the labor costs are normally distributed, it can be shown that the total costs are normally distributed as well; that is, the probability distribution of T is $N(300{,}000, 369{,}000{,}000)$. The probability that the actual cost is equal to greater than $360,000 can be computed by using a standardized normal distribution. Recall, the standard normal random variable Z is computed by subtracting the mean and dividing by standard deviation. Thus, Z in this case is

$$Z = \frac{(360{,}000 - 300{,}000)}{19{,}209} = 3.12$$

From the cumulative probability distribution table for standardized normal, a Z value of 3.12 corresponds to 0.99910. This implies that the cumulative probability, F(z < 3.12) equals 0.99910. Alternatively, the probability that the actual total costs would be $360,000 or higher is 0.00090, or about 0.09% chance. Hence, in this case the contractor has statistical basis to question the supervisor on the excess costs incurred.

8.10 Chapter Summary

This chapter extended the probability concepts to a more pertinent continuous variable setting. Most situations in accounting and auditing relate to large account balances and percentages, both of which are continuous variables. The chapter presented the commonly applied continuous probability distributions of uniform, normal, and exponential. The key concept presented in this chapter is that of cumulative probability distribution. As discussed, in a continuous distribution, the probability of any specific value is negligible and therefore of limited interest. More pertinent is the probability judgment on a range of values. The cumulative probability distributions enable the computation of the probability of a range. The mechanics of computation was illustrated through many numerical examples.

Normal probability distribution was presented as the most common and useful of the various continuous probability distributions. Key features of a normal distribution are that it is symmetrical, thus the mean and median are the same. Normal distribution is also bell-shaped, thus there is higher probability of values around the mean, and the probability decreases as the values are further away from the mean. The two parameters that define a normal distribution are the mean and the standard deviation. Smaller standard deviations imply a narrower but taller distribution, whereas a large standard deviation implies a shorter but wider probability distribution. Any normal distribution can be represented in terms of the standard normal distribution, which has a mean of zero and a variance of one. The transformation of normal distributions to standard normal was extensively discussed. While normal distribution is commonly witnessed in real life, not all data-sets conform to a normal distribution, thus the tests of normality is imperative prior to applying the concepts and the implications of a normal distribution to any data-set. Finally, the mechanics of aggregating two or more random variables that are independent and have a normal distribution was discussed. The concepts presented in this chapter are critical to understanding the subsequent chapters.

Appendix: Using Scientific Calculator to Compute Probabilities

This appendix presents the steps to follow on a scientific calculator (such as TI-84) to find cumulative probabilities for a normal distribution given the z-score. This is illustrated by using the numerical examples presented in the text. Recall that the sales were normally distributed with a mean of 600 gallons and a standard deviation of 75 gallons.

a. The probability that the sales are less than 560 gallons.

$X = 560$, $\mu = 600$ and $\sigma = \sqrt{5625} = 75$

Thus, $\dfrac{(560-600)}{75} = -0.5333$

Using the TI-84 scientific calculator the corresponding probability can be obtained as follows:

- Press the function "2ND" followed by "VARS" (DISTR).
- A menu appears on the screen, select the second by pressing "2". It is "normalcdf("
- Enter "–1E99, -0.5333)". The "–1E99" refers to $-\infty$ and -0.5333 is the z-value. Hence, the cumulative distribution function is computed from a lower limit of $-\infty$ to an upper limit of -0.5333.
- Press "Enter" to obtain the result of 0.2969.

b. The probability that the sales are greater than 720 gallons.

$X = 720$, $\mu = 600$ and $\sigma = \sqrt{5625} = 75$

Thus, $z = \dfrac{(720-600)}{75} = 1.6000$

- Press the function "2ND" followed by "VARS" (DISTR).
- A menu appears on the screen, select the second by pressing "2". It is "normalcdf("

- Enter "1.6, 1E99)". The "1E99" refers to +∞ and 1.6 is the z-value. Hence, the cumulative distribution function is computed from a lower limit of 1.6 to an upper limit of +∞.
- Press "Enter" to obtain the result of 0.0548.

c. The probability that the sales are between 580 gallons and 700 gallons.

$X_1 = 580$, $X_2 = 700$ $\mu = 600$ and $\sigma = \sqrt{5625} = 75$

Thus, $z_1 = \dfrac{(580-600)}{75} = -0.2666$

and $z_2 = \dfrac{(700-600)}{75} = 1.333$

- Press the function "2ND" followed by "VARS" (DISTR).
- A menu appears on the screen, select the second by pressing "2". It is "normalcdf("
- Enter "-0.2666, 1.33)". The cumulative distribution function is computed from a lower limit of -0.2666 to an upper limit of 1.33.
- Press "Enter" to obtain the result of 0.5139.

d. There is a 10% probability that the sales will be less than X gallons.

$\mu = 600$, $\sigma = \sqrt{5625}$ or 75, probability is 0.1

- Press the function "2ND" followed by "VARS" (DISTR).
- A menu appears on the screen, select the third by pressing "3". It is "invNorm("
- Enter "0.1)".
- Press "Enter" to obtain the result of −1.28155. This is the z-value.
- Solve the equation 600 − 1.28 (75) and obtain 504 gallons as the answer.

9

Sampling Theory and Techniques

> *Managers will take little buckets and row their boats over this ocean of facts, dipping the buckets here and there. By examining what they find in the buckets, they are able to draw firm conclusions not only about what is in the buckets, but also about the entire ocean.*
> —George S. Odiorne (1969)

9.1 Introduction

This opening quote, though poetic, is an accurate description of the sampling methodology and its general use in business decision making. In a financial context the auditor and forensic accountant do encounter an "ocean" of data and facts. As accountants might not be able to verify, attest to, or utilize all the information that is potentially available, they must selectively focus on only some of the available information. This process is known as *sampling*.

Sampling is widely used in situations where complete enumeration of data is not feasible. Sampling procedures are employed for their effectiveness, cost, timeliness, and other advantages. The use of sampling rather than censuses for purposes of timeliness occurs in many significant areas in macroeconomics. For example, the government data on unemployment statistics, prices, and inflation are collected on a sample basis at periodic intervals. In these situations, timeliness of the information is of considerable importance. The use of proper sampling technique ensures that the data is collected efficiently without compromising the effectiveness of the information.

Sampling applications for forensic accounting and auditing differs from its applications for other business disciplines. Most business applications of sampling involve decisions where the event will occur in the future and the necessary information is not available. This is primarily the situation in marketing of a new product where the relevant information required is the potential market penetration of the product. Capital budgeting is another example in which decision is based on anticipated future events. In the forensic accounting and auditing context, often complete information is available, but it can require excessive time and effort to process. In such situations a properly selected sample can provide scientific inference on the population as a whole. The distinction in the motivation for sampling between different business applications is an important one. Although for forward-looking events, such as in finance and marketing, the sampling procedure is necessary because relevant information is not readily available for analyzing the events that occurred in the past. In forensic accounting, too much information is available, hence sampling techniques are employed for efficiency.

This chapter begins by discussing the motivation for sampling. Next the theoretical basis for sampling is discussed, and various statistical sampling techniques as well as non-statistical sampling techniques are enumerated. Next comes a discussion on the common sampling approaches applied in auditing and accounting. The chapter concludes with a brief summary.

9.2 Motivation for Sampling

Often times in forensic accounting the auditor will have access to the entire relevant population. When each item of the population is examined or measured, the set of measurement for the entire population is called a census. If the population is relatively small or the information required is extremely critical, a census should be used. However, in many accounting situations a census may not be optimal. Even when census is possible, sampling is often used.

Sampling takes much less time than conducting a census. For instance, in a marketing survey a sample of 100 individuals is much more quickly conducted than a census of 1,000,000. One of the key reasons for using samples in a forensic accounting context is the promptness with which sampling can be administered compared to taking a census. Prompt and timely information obtained from sampling could lead to quick implementation of remedial measures, should they be necessary. Even for macro-economic variables, sampling is used when the information needs are time-sensitive. For instance employment data is released promptly at the end of the quarter, hence the information is generated through sampling. In contrast, U.S. demographic information is obtained through census conducted once every ten years, and the data released several years after the census has been conducted.

The cost of taking the census could be prohibitive. The U.S. Census Bureau is estimated to have spent a few billion dollars in conducting the census on the U.S. population. A forensic accountant may not have resources available to conduct a full census of financial data in a situation where the likelihood of fraud is remote. The cost of gathering the information should not exceed the benefit obtained from the information.

Counter-intuitive as it might appear, the results of a sample can be more accurate than the results of a census. This is because of fatigue and boredom that may set in for the person conducting the census. An individual can be more focused and thorough when gathering data from fewer sources. There are likely to be fewer human errors in measurement and analysis on a survey than on a census.

Additionally, sometimes taking a census is practically infeasible. In a forensic accounting setting, conducting a physical inventory verification of each and every item at each and every location simultaneously is not practical. Further, in certain situations, legal restrictions and privacy of information may inhibit collection of census information.

Even though sampling has distinct advantage over a census, it does not imply that a census should never be used. In forensic accounting situations when the population size is relatively small, it may be worthwhile to conduct a census rather than resort to sampling. Further, because most accounting data is currently stored electronically and some testing is automated, a census is easy to conduct. Thus a census should not automatically be eliminated as a possible means to gather information. Each situation is different, and these factors have to be weighed for each specific instance to determine whether sampling or census is the optimal means to gather information.

9.3 Theory Behind Sampling

Items can be selected in a variety of ways to form a sample. Some of these have human judgment involved, and some don't. The sampling techniques that involve some form of human interference are termed *nonrandom* or *judgment sampling* methods. On the other hand, sampling techniques free from human interference are termed *random sampling* methods, also known as *statistical sampling* or *probability sampling*. The sampling technique could also involve the use of both methods sequentially. That is, first a random sampling method is used followed by judgment sampling, or vice versa.

Judgment sampling or nonrandom sampling involves human interference at some part of the process, hence the items selected are subjective. On the other hand, random sampling does not involve any human judgment; the process is mechanical, thereby preserving the objectivity of items selected. As the results of judgment sampling indirectly depend on personal judgment, the objectivity is compromised, and the reliability of the results cannot be ascertained using objective means of data analysis.

In random sampling, data items are selected by chance alone once the probabilities of selecting the items in the population are known. The individual selecting the sample does not affect those probabilities. This is a critical advantage of random sampling over judgment sampling due to the objectivity inherent in the data gathering process. In random sampling, the precision or reliability of estimates can be obtained from the sample itself. The random sampling techniques thus provide an objective basis for measuring errors due to the sampling process and for stating the degree of confidence that can be placed on the results.

Some situations might require use of both techniques concurrently. Say, for instance, a survey of individuals is being conducted at 10 locations across town. The locations can be randomly picked, but the individual participants are picked by the interviewer. The latter pick is subjective and is based on the personal preference of the interviewer. The second phase of the sampling can be randomized if the interviewer is instructed to survey, say, the seventh person passing an intersection.

In other situations, judgment sampling could be a precursor to statistical sampling. The results obtained from judgment sampling could help determine the parameters for random sampling. Though the results obtained from judgment sampling are limited only to the sample, the results obtained from random sampling could be extended to the population and the reliability of the estimate as well as confidence in the result can be mathematically computed.

The next two sections discuss commonly used statistical and nonstatistical sampling techniques and assess the usefulness of each in the context of forensic accounting.

9.4 Statistical Sampling Techniques

The most commonly used and widely known sampling technique is *random sampling*. Random sampling is a method of placing items from the population into the sample such that each item of the population has an equal chance of being selected. Random sampling can be conducted for both finite and infinite populations. The major body of statistical theory of inference is based on the simple random sampling.

When selecting a random sample of n from a finite population of N elements, two approaches are possible: *selection with replacement* and *selection without replacement*. When selected items are placed back in the population and may be reselected, then the sampling is with replacement. This is done to ensure that the fundamental condition of random sampling, that each item has the same probability of being selected, is not violated. However, this procedure creates the possibility that the same item could be selected twice or multiple times in the sample. In most accounting and auditing situations, when the intent is to maximize the number of items examined, duplication of the sample items might not be desirable. Therefore, in auditing and accounting contexts, random sampling is often undertaken without replacement.

Sampling without replacement eliminates the possibility of selecting the same item twice in the sample. This violates the theoretical principle underlying random sampling, which is each item should have an equal probability of being selected. When selecting from a population size of N, the first item selected had a probability of $1/_N$. If the selection is without replacement after 10 items have been selected, the remaining population size is (N − 10), thus the probability of a remaining item in the population to be selected as the 11th item is $\frac{1}{N-10}$. As the denominator decreases, the probability increases.

This change in the denominator for probability computation is immaterial when a small sample is selected from a large population. Say, if you were to select a sample of 200 items from a population of 100,000, the probability that a particular item of the population will be selected first is $\frac{1}{100,000}$, or 0.00001, and the probability that of the remaining items one would be selected as the 200th item of the sample is $\frac{1}{99,801}$, or 0.000010019. The increase in probability is 0.000000019, an insignificant amount in accounting and auditing context. However, when a large sample is selected from a small population, the selection of the last item in the sample will have a significantly higher probability than that of the first item selected. To illustrate, say a sample of 80 is selected from a population of 400. The probability that a particular item of the population will be selected first is $\frac{1}{400}$, or 0.0025, and the probability that of the remaining items one would be selected as the 80th item of the sample is $\frac{1}{321}$, or 0.0031. The increase in probability is 0.0006, while still small, is about a 25% increase from the original probability.

A simple random sample can be obtained in a variety of ways. The most common method is to assign a number to each element of the population and then use a random number generator to generate numbers to select items that correspond to the numbers randomly generated. The appendix at the end of the chapter provides instructions on using Excel to generate random numbers. Alternatively, a lottery can be constructed if the population size is manageable.

Systematic sampling is an easily implementable procedure to randomly select items in the sample. In this case the population is shuffled so that the items in the population are randomly designated. A systematic approach to selecting the sample is then outlined, say, select every 17th item. Selecting a sample based on a predetermined system is called systematic sampling. The important aspect to this type of sampling is to be able to shuffle the population so that the items in the population can be randomly ordered. This procedure is most commonly followed in physical surveys where a surveyor interviews, for example, every 26th person leaving the store. A systematic sample is easy to select as it requires no special expertise.

In many accounting situations, the population is heterogeneous. Random selection of a sample from the population might misrepresent a particular segment of the population. In such cases, *stratified random sampling* should be used. The population has to be first stratified or divided into subgroups based on identifiable characteristics of the population. Next, sample items can be randomly selected from each of these subgroups or strata.

Stratified random sampling is especially relevant in an accounting context. Many accounting data-sets are heterogeneous or have high variances. To ensure adequate representation of each important characteristic of the population, stratification may be necessary. Consider for example an audit of outstanding accounts receivables. Usually, an organization's accounts receivables will comprise of few very large balances and numerous small balances. If a random sampling is undertaken for the entire population, the sample would likely consist of only the small accounts because there are many more of them, and the probability of selecting any item is constant across the population. However, if one presumes that large account balances have large errors and small account balances have small errors, predominantly selecting small account balances will underestimate the error in the account. In such situations, stratification of the account based on size will be appropriate.

When conducted properly, stratified sampling can be used to estimate a population's characteristics more accurately than a similar sized random sample. As sampling and subsequent verification is costly, this implies that stratified sampling could be more cost effective than random sampling. Stratification is particularly useful when the population contains extreme outliers. In a forensic accounting and auditing context, such outliers need to be examined by auditors and not excluded from consideration. Random sampling increases the chance that such outliers, if few, will not be represented in the sample. Stratified sampling, on the other hand, ensures inclusion of at least a few of these outliers in the sample to ensure their testing.

Another valid sampling technique of relevance to accountants is *cluster sampling*. This approach is beneficial when the items in the population are geographically dispersed. For instance, inventory warehouses for a global retailer, such as Home Depot or Wal-Mart, are spread all over the world. A random selection of locations to visit may require auditors to visit multiple countries where the retailer has stores or warehouse facilities. This could be costly as well as a time-consuming process. In such situations cluster sampling, a modified random sampling technique, could be used.

Like in stratified sampling, first the items of the population are divided into subgroups based on geographical location. However, unlike in stratified sampling, items from each strata are not represented in the sample; instead, a subgroup is either selected to be audited or not selected at all. If the subgroup is not selected, then none of the items from that subgroup would be included in the sample. The process of cluster sampling is a three-step process: first items are divided into subgroups or clusters; next, a few of these clusters are selected; and finally items from the selected clusters are randomly selected to form the sample.

Consider the following example of multiple inventory locations dispersed all over the globe. The number of locations in various regions is graphed in Figure 9.1. Each box represents a geographic region. The numbers of inventory locations in these regions are written in respective boxes. The regions are disaggregated into smaller and smaller regions as you go down the tree. Using a cluster sampling methodology, some of these clusters of regions are selected. Next, a sample of locations to be visited is randomly selected from the clusters. Figure 9.2 shows the number of locations in the sample.

Just as with stratified sampling, a well-designed cluster sample will generally provide information of comparable quality to random sampling but at a much lower cost. However, to be able to apply cluster sampling you must know that each cluster is representative of the population. If the clusters themselves are different, then selecting samples from a cluster provides generalities regarding only the cluster and not necessarily the entire population. Ideally, the clusters should not be homogeneous; rather, they should exhibit the same heterogeneity as the underlying population. When applying the cluster sampling technique to select inventory locations to visit, the auditor should consider these caveats prior to generalizing the sample results to the population. The data mining technique of clustering introduced in Chapter 5, "Data Mining," can be useful in forming clusters of worldwide inventory locations.

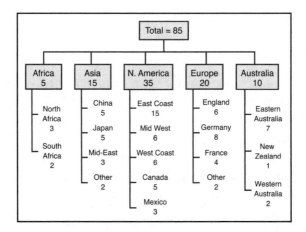

Figure 9.1 Inventory locations for a worldwide retailer

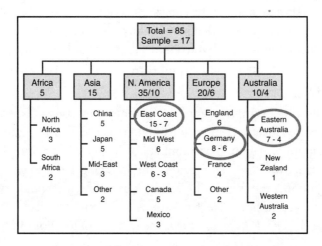

Figure 9.2 Cluster sampling of inventory locations to audit

9.5   Nonstatistical Sampling Techniques

In some situations a probability sample is either not possible or not desirable. Consider for example, truckloads of potatoes coming into a manufacturing plant that makes frozen French fries. A quality control inspector either accepts or rejects the entire truckload of potatoes. The inspector may take a sample of a few potatoes from the truck and test them for quality. A random selection of potatoes to test is not only time-consuming and costly but might be infeasible due to the difficulty of assigning a specific identification number to each potato in the truck. Moreover, such probability-based sampling is perhaps unnecessary. The inspector would visually inspect the shipment and based on his experience would pick a few samples to test. Such nonprobabilistic sampling techniques are commonly used in practice and are known as *judgment sampling*. This method is distinct from the ones described in the previous section in that the person taking the sample has direct or indirect control over what specific items are selected in the sample.

Judgment sampling is also used in cases when potentially some items in the population are more representative of the population than others. This is relevant in marketing campaigns and surveys where rather than randomly selecting the city to survey for a new product launch a representative city is selected based on judgment of the marketing specialists. That is the reason why many marketing surveys and new product launches tend to happen in larger metropolitan cities like New York and Los Angeles rather than in rural towns.

In a forensic accounting context, to test for related party transactions, the auditor might hand-pick a few round-trip transactions[1] that she finds suspicious. In such cases, rather than randomly sampling through all transactions for validity, the auditor uses her judgment to select a few items. Clearly, in such cases the sample selected is subjective and bears characteristics of the person selecting the sample. Consequently, the sample is only as good as the person doing the sampling. If poor judgment is used, the information gathered might not be representative, hence potentially misleading. A surveyor can easily bias an opinion survey by selecting individuals who share similar views. For example, a TV channel might survey only its viewers and present the results of such a survey as that of the general population.

Some situations in forensic accounting require the use of judgment or other nonstatistical sampling techniques. Although such use is permissible, it is important to remember that the sample results cannot be statistically analyzed. Hence, when judgment sampling is used, the results are valid only for the sample and cannot be projected to the population that was not part of the sample. Therefore, in forensic accounting cases where statistical evidence is presented, it is imperative to assess the validity of the sampling process to ensure whether the statistical inferences are tenable. More importantly, because judgment sampling is subjective, the quality of the sample is affected by the person undertaking the sampling. Hence, when evaluating evidence obtained from judgment sampling, it is important to verify the qualifications, motives, and affiliation of the person selecting the sample. The biases and motives of the individual have to be taken into consideration.

Despite the caveats of judgment sampling just discussed, it has its uses in forensic accounting situations if undertaken properly. First, the investigator or the forensic accountant has to be in charge of selecting the sample and should be the one exercising his or her judgment. It is important for the investigator to ensure that they are not being influenced in their judgment by individuals who could potentially be perpetrators. Further, since the results from judgment sampling cannot be extended to the population at large, the objective of judgment sampling in a forensic accounting context is to unravel "smoking gun" evidence. See Exhibit 9.1 for the "smoking gun" evidence uncovered in Enron's demise, known as Enron's Nigerian Barge Deal. The forensic accountant should be encouraged to use their experience and expertise in identifying suspicious items to further investigate, formally known as judgment sampling. However, they have to be careful so as to not to be unduly influenced while selecting those items or overgeneralize their findings based on judgment sampling to the general population.

Exhibit 9.1: Enron's Nigerian Barge Deal

One of Enron's thousands of questionable transactions was considered critically important at the time because of the impact it had not only on Enron but on how Wall Street firms structure their deals. It was also important because of the "smoking gun" evidence that the case presented, which led to conviction of two mid-level Enron employees and four bankers from Merrill Lynch. Though the profits from the deal were $12 million, paltry compared to Enron's size, it was one of the first cases tried by the prosecutors against Enron officials.

The events related to the purchase and sale of barges originated from a deal that Enron had with the Nigerian government of building a power plant near Lagos. The total cost of the project was estimated to be $500 million. Although the power plant was being built, Enron committed to supply power from barge-mounted power plants. The barges were to start supplying power as early as the fall of 1999 even though the construction of the power plants wasn't set to begin until the latter part of 2000. Consequently, Enron had purchased the barges for a considerable sum.

However, by early 2000 before any construction had begun, the project was facing many political hurdles. The contracts were being scrutinized by World Bank and other authorities in Nigeria. Moreover, the cost estimate was revised to upward of $800 million. Facing such uncertainties and huge losses, Enron started selling parts of the deal to Merrill Lynch, which turned around and sold those to a shell company related to Enron. The first sale to Merrill and Merrill's quick sale of its interest back was at the heart of the criminal allegation of a "round-trip" transaction. So while Enron recorded profits from its transaction in its books, the obligation and losses were hidden in the books of its related but unconsolidated entity. Merrill took a considerable fee of about $17 million for structuring and participating in the transaction. The SEC claimed that because Merrill's executives knew that Enron would fraudulently report sale and profits from this deal, their passive participation is tantamount to aiding and abetting Enron's earning manipulation.

> Although the piece-meal portions sold to Merrill Lynch amounted to more than $225 million, the Nigerian barges, being tangible and easily traceable, led to the "smoking gun" evidence for the prosecutors. Selection of those two transactions to audit out of millions of transactions that Enron and Merrill engaged with each other is attributable to the instinct and experience of the forensic accountant who, while employing judgment sampling to pick transactions to investigate, picked those two.
>
> Merrill Lynch had received an exorbitant fee of $17 million for entering into the transaction. The SEC charged that Merrill believed that it was getting into a "round-trip" transaction and had prearranged the sale prior to entering into the purchase transaction and treated the two transactions as a "wash." The SEC further alleged that Merrill knew, or should have known, that it would have a significant impact on Enron's reported results. Merrill Lynch for its part fired the offending executives and settled with the SEC by paying fines of $80 million but without admitting or denying fault.[2]

9.6 Sampling Approaches in Auditing

AU Section 350 provides authoritative guidance on audit sampling.[3] It acknowledges the use of both probabilistic and nonprobabilistic methods of sampling, and its guidance is applicable to both methods. The guidance asserts that the basic concept of sampling and the risk inherent with the approach is well established in audit practice (paragraph 07) and distinguishes between sampling risks and nonsampling risks. It defines sampling risk as the risk that the auditor assumes in generalizing the sample results to the population. This risk arises from the chance that "a particular sample may contain proportionately more or less monetary mis-statements" and notes that sampling risk varies inversely with sample size. That is, the greater the sample size, the smaller the risk and the smaller the sample size, the greater the risk.

Paragraph 17 of the Audit Sampling guidance (AU 350) requires auditors to consider specific audit objectives in determining what population to draw the sample from. For example, when testing for misappropriation of funds from cash sales, the sample should not be drawn from recorded sales, rather from merchandise sold or shipped. When the allowable risk of incorrect acceptance increases, the sample size for the associated audit test could decrease. Paragraph 22 implicitly allows the usage of both stratified sampling and cluster sampling in appropriate audit situations. It states "auditor may be able to reduce the required sample size by separating items into homogeneous groups..."

In a forensic accounting and auditing context, there are many attributes of an account that could be subject to testing. It is imperative that the auditor clearly defines which attribute needs to be tested and designs the sampling plan accordingly. For example, if the forensic accountant suspects fraud in the Purchasing/Payable department in which payments are being made by the organization for goods not received or ordered, there are three source documents that the auditor could sample from: receiving records, purchase orders, and disbursement vouchers. A sample selected from receiving records matched to the purchase order and disbursement vouchers will not be able to detect such a fraud. Likewise, if payments are made without requisite purchase orders, a sample obtained from the purchase order verified for payments and receipt of goods will fail to detect such fraud. The sample has to be taken from the disbursement records and matched whether goods were received and/or whether a legitimate purchase order exists.

Usually there are multiple attributes of an account that may be of interest to the forensic accountant; however, there are very few audit tests and sampling methods that could provide evidence relevant for a comprehensive set of attributes. Thus, a single sample to test for all relevant attributes of an account may be inadequate. In such situations either multiple samples are gathered using a different sampling design or the most critical of the attributes is tested. The sampling plan, and hence the sample, in an auditing/forensic accounting context is based on a specific attribute. It is important for the auditor to specify the attribute unambiguously prior to designing a sampling plan or selecting a sample.

There are four additional sampling techniques unique to accounting/auditing. These are: discovery sampling, acceptance sampling, PPS sampling or probability proportional to size, and DUS also known as the dollar unit sampling. Each of the four methods is briefly described in the following paragraphs.[4]

Discovery sampling is an ultra-conservative approach to interpreting sample results. It requires rejection of the entire population if even one error is detected in the sample. It is a conservative approach as it minimizes the risk of wrongly accepting a mis-stated population. This approach is appropriate in high risk audits or scenarios in which detection of even one error is sufficient cause to reject the entire population. This approach implicitly assumes that there is a zero error rate in the population, hence even one detected error in the sample is contradictory to that premise. Due to its conservative approach to error, the auditor using discovery sampling is likely to over-audit in situations when there is non-zero yet negligible error rate in the population.

Acceptance sampling is a bit more tolerant than discovery sampling in that it allows for few errors to be present in the sample and still accepts the population. In this sampling method, a low number of errors are accepted, if errors detected in the sample are higher than the prespecified number, as with discovery sampling, the entire population is rejected. It is less conservative than discovery sampling in that it allows for a few errors to be detected in the sample. When an acceptable number of errors in acceptance sampling is reduced to zero, the method reverts to discovery sampling. It is important to note that both discovery and acceptance sampling require no further computations or use of statistical methodology after the sample results are known. In discovery sampling the population is accepted if no errors are detected in the sample and rejected if even one error is detected. Likewise in acceptance sampling the population is accepted if the number of errors detected in the sample is less than the ex-ante number, else the population is rejected.

An adaptation of stratified sampling discussed earlier is *probability proportional to size* sampling, or PPS sampling. As evident from its name, larger account balances have a higher probability of being selected in the sample than smaller account balances. As the resultant sample is likely to have more items with higher balances, the auditors' efforts on detecting large errors get focused on the larger accounts. Though it may appear that a selection method such as PPS violates the equal probability of each item assumption of probabilistic sampling, it does not. In this case, the dollar amount is the unit of interest and not the individual account. When the unit is defined as such, the probability that any unit would be selected in the sample remains the same throughout the population. The sample size obtained using a PPS sampling method is often smaller than the equivalent sample size that would be obtained through a random sampling method. *Dollar unit sampling* (DUS) is an implementation approach to the PPS sampling technique. In DUS, each dollar amount is considered a unit and is assigned a random number. When a particular dollar amount is selected, the corresponding account is included in the sample. As large account balances have more dollars represented in the population than small account balances, these have a higher chance of being included in the sample. Although the dollar unit sampling technique achieves the objective of PPS, it is not based on normal distribution assumptions, and hence the projection of the sample results could be difficult.

9.7 Chapter Summary

This chapter presented the theory of sampling and provided a description of commonly used and widely accepted sampling techniques. One of the important ex-ante decisions that the forensic accountant or the auditor has to make is whether they wish to generalize the sample results to the entire population. The choice of the sampling technique might be restricted based on whether the sample results will be generalized to the population or not. When generalization is desired, the sampling technique has to be probabilistic and the sample randomly selected. When such generalization of sample result is not required, the sampling techniques could be judgmental or nonprobabilistic. A combination of two types of sampling methodology is also permissible but restricts the ability to generalize the results.

The key determination is the primary intent of testing or audit. In certain forensic accounting engagements, the objective is to detect fraud, and hence an evidence of a "smoking gun" is desirable. In such a forensic accounting engagement, the accountant need not provide assurance on absence of fraud. For these a judgment sampling that leverages on the experience, expertise, and skills of the investigator has its advantages. However, in more traditional auditing applications the audit objective is to provide assurance that the account balances are not materially misstated. In such cases, the importance of a "smoking gun" is diminished; instead, the ability to generalize sample results is a key objective. For such cases, the use of probability sampling techniques is recommended.

Appendix: Random Number Generation in Excel

Below are the steps to generate random numbers using Excel.

1. On the menu ribbon, click on the tab labeled "Data".
2. Click on the "Data Analysis" icon on the right side of the menu bar. If the icon does not appear, it might have to be activated by going to the "office" menu on the upper left corner. (Steps to activate the "Data Analysis" tool are provided in the Appendix to Chapter 12.)
3. A pop up window appears with the list of Analysis Tools.
4. Scroll down to "Random Number Generation", highlight it, and click OK.
5. Another pop up window appears.
6. Fill in the details. Use "Discrete" for Distribution.
7. In the box under "Parameters", enter the range from which items will be selected.
8. Click OK, the requisite number of random numbers will appear on the worksheet.

Additionally, websites such as http://www.mathgoodies.com/calculators/random_no_custom.html provide calculators that generate random numbers for a specified range.

9.8 Endnotes

1. This is a common terminology for transactions that organizations undertake with another in an accounting period that are reversed in the subsequent accounting period. For example, Organization A sells high priced items to B in 2000. In 2001, B sells those items back to A at a predetermined price. Repo 105 transaction undertaken by Lehman, and discussed previously in the book, is an example of round-trip transactions.

2. The SEC News Release on the settlement with Merrill released on March 17, 2003 is available at http://www.sec.gov/news/press/2003-32.htm.

3. The material in this chapter is supplemental to the Audit Sampling guidance from the PCAOB available at http://pcaobus.org/Standards/Auditing/Pages/AU350.aspx., and similar guidance from IFAC available at http://www.ifac.org/sites/default/files/downloads/a027-2010-iaasb-handbook-isa-530.pdf. The purpose in this chapter was not to replicate or summarize the guidance, but to provide the statistical foundation underlying them.

4. For a detailed description of the methods see *Statistical Auditing: Review, concepts and problems*, by Andrew D. Bailey Jr. Harcourt, Brace Jovanovich Inc. 1981.

10

Statistical Inference from Sample Information

It is a capital offense to theorize before one has the data.
—Sir Arthur Conan Doyle (Sherlock Holmes)

10.1 Introduction

Data-sets in accounting are extremely large, hence time and cost constraints make it impossible to test all the items. This requires the use of sampling. The process of appropriately selecting a sample was discussed in the previous chapter. When the sample is selected using a parametric or probabilistic approach, the results of the sample could be generalized to the population. This chapter discusses theory and methodology for these generalizations.

In most forensic accounting and auditing settings, a relevant sample of the population is selected and tested. The key metrics are computed for the sample, and the attempt is to infer the population parameters from the observations made on the sample. The auditor selects and tests a specific sample and documents the findings for the sample. A statistician is concerned about how many possible samples could have been hypothetically selected, using the auditor's selection criteria, in generalizing the sample result to the population.

Information obtained from a sample can be perceived as an incomplete observation regarding the population characteristics. Basing judgments on incomplete information introduces uncertainty and risks. In such cases, it may not suffice to merely know the judgment but also the extent of uncertainty and risks underlying those judgments. Statistical inference is a methodology that provides a theoretical basis for making reasonable decisions or judgments from incomplete information. This chapter discusses inference on population characteristics from information contained in samples. It is important to note that concepts discussed in this chapter are valid only when the sample was selected using parametric or probabilistic techniques. In instances when the sample is selected using judgment or other non-parametric techniques, observations from the sample cannot be generalized to the population using the approaches presented in this chapter.

The chapter begins by presenting the statistical concepts underlying distribution of the sample mean, which leads to the estimation of sampling error. This theory is the essential building block to understanding statistical estimation of the population parameters and the construction confidence intervals discussed later in the chapter. Next comes a discussion of an important result of statistics known as the Central Limit Theorem, which allows us to use the properties of normal distribution to infer from the sample mean even when the underlying population may not be normally distributed. The point estimate and the desirable properties of the estimator are discussed next. The chapter concludes with confidence intervals and the equations and the procedures for constructing confidence intervals for various scenarios.

10.2 The Ability to Generalize Sample Data to Population Parameters

In addition to ensuring that the sample is properly tested and the resultant data is properly analyzed and presented, the forensic accountant or auditor faces a more challenging task when sampling from a population. This challenge stems from the fact that two samples taken from the same population would lead to different sample values, hence potentially lead to different and possibly contradictory inferences. This is an important concept for the auditor and accountants to understand and is illustrated with an example. Consider a population consisting of 40 numbers shown in Table 10.1. The task is to estimate the population mean using a sample of five items. For the purpose of illustration there are two samples of five items: Sample 1 consists of items in Column 5; Sample 2 consists of items in the prime diagonal that is cell 11, 22, 33, 44, and 55. The five items in Sample 1 are {37, 44, 42, 40 and 33}, and those in Sample 2 are {30, 36, 15, 28, and 33}. You might want to select your own sample of five items and repeat the steps of the analysis here.

Table 10.1 A Population of 40 Items for Illustration

Column 1	Column 2	Column 3	Column 4	Column 5	Column 6	Column 7	Column 8
30	22	18	45	37	26	42	12
25	36	33	47	44	26	20	34
28	46	15	41	42	29	16	37
31	30	48	28	40	22	18	42
21	17	29	36	33	47	25	31

Mean = = 31.225
Variance = σ^2 = 102.076
Standard Deviation = σ = 10.1

With the five data items, the mean in Sample 1 is 39.2, and standard deviation is 4.32, while the mean of Sample 2 is 28.4 and standard deviation is 8.08. Clearly neither of the sample means is equal to the population mean of 31.225. Although two randomly selected samples are used, there are 658,008 different samples of five items that can be taken from a population of 40 items. This can be computed as

$$\frac{40}{5} = \frac{40!}{5!35!} = 658,008$$

Each one of these 658,008 possible samples would have a mean and a standard deviation that are very likely different from the population mean and standard deviation these are trying to estimate. While selecting a sample of five items from the population it is possible, though remote, to select the five largest items (Sample 3) or the smallest five items (Sample 4). For Sample 3, the five items are 48, 47, 47, 46 and 45 which has a mean of 46.6 and a standard deviation of 1.14. For Sample 4, the five items are 12, 15, 16, 17, and 18, which has a mean of 15.6 and a standard deviation of 2.30. Clearly, if you were to select either Sample 3 or Sample 4 you would make incorrect inferences for the population characteristics. Fortunately, the probability of selecting Sample 3 is one in 658,008, or about 0.0000015. This is an extremely remote possibility, but a possibility nonetheless and hence has to be considered.[1]

The difference between a population value and the corresponding sample value is called the *sampling error*. The exact amount of sampling error is determined by which sample is chosen and the mean of that sample. As you do not know how large the sampling error would be before selecting the sample, the knowledge of the distribution of sample mean is useful. The theoretical frequency distribution of the sample mean of all 658,008 samples of five items would constitute the sampling distribution of the means for samples of five items. The probability distribution of the sample means is integral to making statistical inferences on the population based on sample results.

As discussed in previous chapters, a probability distribution can be described with three attributes: its type, mean, and standard deviation. These characteristics can be mathematically derived, but here we intuitively discuss the characteristics of these distributions. A sample mean is just as likely to be above population mean as below it. In the previous example of two samples, one had a mean greater than the population mean, and the other was lower than the population mean. Further, small deviations from the population mean will occur more frequently than large deviations from it. Additionally, due to averaging in the sample you would expect the sample means to be less dispersed than the items in the population. In summary, expect the distribution of the sample means to be symmetric, centered around the population mean but with a lower standard deviation than the original population.

When the size of the sample is altered the probability distribution of the sample mean is expected to change. Although the symmetrical property and the central tendency of the distribution remain unchanged, the standard deviation is expected to decrease as the sample size increases. At the extreme, when sample size equals the population size, there is only one sample possible, and hence there is no dispersion around the mean. In other words, as sample size increases, the standard deviation of the sampling distribution decreases. Figure 10.1 illustrates the relationships. While the x-axis denotes values of individual items for the population distribution, the x-axis denotes the value of the sample mean for the sampling distribution. As all three distributions are continuous distribution, probabilities are represented by computing the area under respective curves.

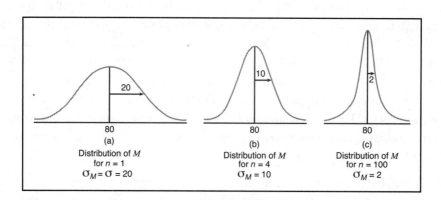

Figure 10.1   Distribution of Sample Means

A critical mathematical result, which is not intuitively obvious, is that when the original population has normal distribution, the sample means are also distributed along a normal curve. That is, the sampling distributions are also normally distributed. This mathematical result has important implications. Based on this theorem it can be deduced that the standard deviation of the sampling distribution becomes a measure of average sampling error.

In summary, the important properties of the sampling distribution of the mean when the population is normally distributed with mean µ and standard deviation θ are

- It is normally distributed.
- It has a mean equal to the population mean.
- Its standard deviation equals the population's standard deviation divided by the square root of the sample size. That is,

$$\text{Sample standard deviation} = \frac{\sigma}{\sqrt{n}}$$

The sample standard deviation is a critical concept. The sample standard deviation is inversely proportional to the sample size, hence increasing sample size decreases the sample standard deviation. As just discussed, the sample standard deviation is a measure of average sampling error, once it decreases due to increase in sample size the sampling error consequently decreases. In other words, the precision with which the sample mean estimates the population mean increases with a reduction in sample standard deviation. As the sample standard deviation is inversely related to both the precision of the estimate and sample size, the latter two terms are directly related. That is, the precision of the estimate increases as the sample size increases. Further, the square root term on the sample size implies diminishing return on sampling effort. That is, increasing the sample size from 1 to 4 achieves the same effect as increasing the sample size from 4 to 16 or from 16 to 64.

10.3 Central Limit Theorem and non-Normal Distributions

Many probability distributions might not be normally distributed. The mathematical results discussed in the earlier sections were based on the underlying assumption that the population was normally distributed. Hence, when the population is not normally distributed the results just discussed have to be revisited. Fortunately, it is a remarkable statistical finding that irrespective of the nature of population distribution, the sample mean is approximately normal for a sufficiently large sample size. This relationship between the population distribution and its sample mean is best summarized in what is the most important theorem of statistical inference, the Central Limit Theorem.

> **Central Limit Theorem**
>
> Consider a random sample of n observations selected from any population with mean μ and standard deviation σ. When n is sufficiently large, the distribution of the sample mean is approximately a normal distribution with mean equal to μ and standard deviation equal to $\frac{\sigma}{\sqrt{n}}$. The larger the sample size, the better the approximation to normal distribution.

The Central Limit Theorem provides much needed assurance that irrespective of the shape of the population distribution, the probability distribution of the sample mean approaches normality as the sample size increases. The key point is that for a wide variety of population types, samples do not have to be very large for the distribution of sample mean to approach normality. Figure 10.2 shows the distribution of the sample mean for different population distributions and different sample sizes. As can be seen, the more skewed the population, the higher the sample size required for the sample mean to have a normal distribution. Although symmetric distributions require a much smaller size, as low as 5, to yield a normal distribution for sample mean, highly skewed distributions require a sample size of 20 to yield a normal distribution of the mean. The commonly practiced rule of thumb mentioned in most statistics textbooks is that a sample size greater than 30 ensures that the sample mean is normally distributed. It is important to note that a sample size of 30 only ensures normal distribution for its mean but in no way claims to be an adequate sample size for various sampling problems. The computation of adequate sample size is discussed in the next chapter.

10.4 Estimation of Population Parameter

Statistical estimation procedures provide the means of obtaining estimates of population parameters with desired degrees of precision. These are useful in situations where the cost of conducting a census of the population is prohibitive or a complete enumeration is impossible as in the case of infinite population. Sampling technique is employed when exact results are not required; instead, estimates derived from sample data provide necessary information.

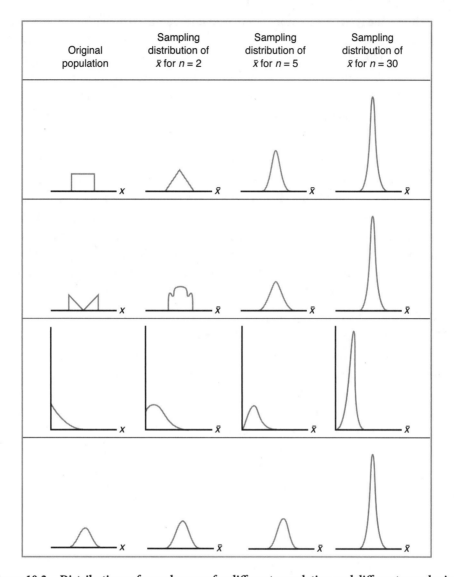

Figure 10.2 Distributions of sample mean for different population and different sample sizes

McClave, James T.; Benson, P. George; Sincich, Terry L.; *A First Course in Business Statistics*, 8th Ed., ©2001, p. 225. Reprinted and electronically reproduced by permission of Pearson Education, Inc., Upper Saddle River, New Jersey.

Considering that estimates are made, some criteria is needed to evaluate the quality of the estimation process, or the estimator. These criteria are unbiasedness, consistency, efficiency, and sufficiency. Although a forensic accountant does not have to evaluate these on a case by case basis, a familiarity with these concepts will help them in proper evaluation of statistical evidence. Each of these is discussed briefly here:

- **Unbiasedness:** An estimator is unbiased if its average value equals the population parameter. An unbiased estimator is correct on an average. This does not imply that it will be correct all of the time or in any particular instance. Recall from the previous section that sampling error is the difference between the sample mean and the population mean. Formally stated, unbiasedness implies that on an average the sampling error will be zero. As shown in the previous section, the sample mean is an unbiased estimator of the population mean.

- **Consistency:** Although unbiasedness implies that the average sampling error is zero, it says nothing about the proximity of individual estimates to the parameter. The criterion of consistency recognizes the need for a small sampling error. An estimator is considered consistent when a larger sample size reduces sampling error. In other words, if the estimation process gets more precise as sample size increases, the process is consistent. Again, as shown in the previous section, increasing sample size reduced standard deviation of the mean of the sample, hence mean is a consistent estimator of the population parameter.

- **Efficiency:** The estimator that leads to lesser variability of the estimate is known as the efficient estimator. When selecting a sample, you could use the mean of the observations or the median to assess the population parameter. Thus far sampling has only been discussed in context of the mean and not the median because as shown in Figure 10.3, the mean is more efficient than the median in estimating the population parameter. The distribution that is less spread out indicates less variability and hence is more efficient.

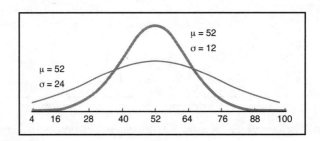

Figure 10.3 Illustration of efficiency of a statistical estimator

- **Sufficiency:** A sufficient estimator uses all the information available in the sample data. The sample mean is a sufficient measure because it uses the value of each sample item.

Two different types of estimates of population parameter are commonly used: the point estimate and the interval estimate. A *point estimate* is a single number used as an estimate of the unknown population parameter. For example, the mean prices of a sample of houses listed for sale in a town is a point estimate of the average price of all houses in that town. A point estimate, although precise, will in most cases not be exactly equal to the population parameter. Hence it is useful and perhaps necessary to have some notion of the degree of error or spread that might be involved in using the point estimate. In those cases an interval estimate of the population parameter might be of importance. An interval estimate states two values within which the population parameter is expected to reside. Thus, the interval estimate of the house prices would give two values within which the average housing price of the town is expected to lie. In most situations, the range of values of the population parameter could be of more use than a single estimate of the population parameter.

In addition to specifying an interval, it may be additionally informative to specify the level of confidence the sample information provides on the interval. Intuitively, the confidence in a particular interval increases as the width of the interval increases. That is, a high degree of confidence can be placed that the population parameter falls within a wider interval than within a narrower interval. The statistical procedure to address this concern is termed confidence interval estimation. The confidence interval is an interval estimate of the population parameter and is discussed in the following section.

10.5 Confidence Intervals

The interval estimate, popularly known as the confidence interval, specifies a range within which the population mean is expected to be. The interval is symmetric around the point estimate, that is the lower limit and the upper limit of the interval are equally apart from the point estimate. There are two main scenarios to consider. In the first scenario the sample is drawn from a normally distributed population with a mean. The sample mean in this case is a normal distribution with a mean of μ and a standard deviation of $\frac{\sigma}{\sqrt{n}}$. In the second scenario, the population is not normally distributed or its distribution is unknown but has a mean of μ and a standard deviation of σ. For the latter cases central limit theorem can be invoked if sample size is greater than 30 and the sampling distribution can be approximated by a normal distribution with a mean of μ and a standard deviation of $\frac{\sigma}{\sqrt{n}}$ as in the first scenario. However, in the first scenario, sample mean of any sized sample would have the property, but in the second scenario the approximation is valid only for sample sizes greater than 30. Also in the second scenario as the sample size increases the approximation gets better.

Recall from Chapter 8, "Continuous Probability Distribution," cumulative probability can be computed by solving the integral for area under the curve. For all practical purposes solving the integral is not required; instead, any normal distribution is transformed to a standard normal by subtracting the mean and dividing by the standard deviation. For a standard normal distribution the area under the curve or z-score can be obtained from tables, scientific calculators, or Excel. A z-score of -1.96 corresponds to cumulative probability of 0.025, and a z-score of +1.96 corresponds to cumulative probability of 0.975. Thus as shown in Figure 10.4 the area between the lower bound of -1.96 and upper bound of +1.96 is 0.95. This provides the theoretical basis for constructing a confidence interval. In the equations shown next, σ is used to generically represent a measure of dispersion around the mean. This measure is discussed later in this section in greater details for specific scenarios, such as large sample sizes and whether the parameter σ is known or has to be estimated.

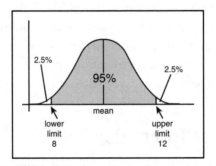

Figure 10.4 Pictorial illustration of a 95% confidence interval

Specifically, 95% of all sample means lie in this range:

$$\mu - 1.96 \frac{\sigma}{\sqrt{n}} \leq x \leq \mu + 1.96 \frac{\sigma}{\sqrt{n}}$$

Likewise 90% of all sample means lie in this range:

$$\mu - 1.645 \frac{\sigma}{\sqrt{n}} \leq x \leq \mu + 1.645 \frac{\sigma}{\sqrt{n}}$$

as the z-score of -1.645 corresponds to a cumulative probability of 0.05 and the z-score of 1.645 corresponds to a cumulative probability of 0.95.

Also, 99% of all sample means lie in the following range:

$$\mu - 2.57 \frac{\sigma}{\sqrt{n}} \leq x \leq \mu + 2.57 \frac{\sigma}{\sqrt{n}}$$

In general, the interval within which the sample means lie can be computed using this generic formula:

> Generic Formulation of Confidence Interval
>
> Point Estimate ± (interval coefficient) (measure of dispersion)

When constructing a confidence interval the user has to specify the desired confidence level on the interval. This confidence level is then transformed into the interval coefficient subject to the underlying assumptions. The measure of dispersion likewise varies based on the assumptions. The following sections describe various scenarios and the corresponding measures of critical value as well as the measure of dispersion. First the scenarios with large sample sizes are covered, followed by scenarios with small sample sizes.

10.6 Confidence Interval for a Large Sample When Population Standard Deviation Is Known

Sample means for a large sample, according to the Central Limit Theorem, are normally distributed regardless of the underlying distribution of the population. As discussed before, any normal distribution can be transformed to a standard normal distribution by subtracting the mean and dividing by the standard deviation. Hence the z-scores from the standard normal tables can be used as the requisite interval coefficient in the computation of confidence intervals. The interval coefficient is the number of standard errors, or z-scores symmetric on either side of the mean to allow the area under the curve to equal the specified confidence level as shown in Figures 10.4 and 10.5.

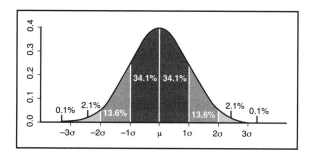

Figure 10.5   Confidence intervals corresponding to 1, 2, and 3 standard deviations

Knowledge of the population standard deviation, denoted as σ_x, is the best case scenario. This parameter can be directly substituted as the measure of dispersion and does not have to be estimated from the sample. The larger the population standard deviation, the wider the confidence interval and lower the precision of estimating the population mean.

For finite population size if the sample is large and forms a significant proportion of the population, a finite correction factor is used. The finite correction factor impacts the measure of dispersion and is multiplied to the standard deviation of the population. The finite correction factor is given as

$$\sqrt{\frac{(N-n)}{(N-1)}}$$

where N is the population size

n is the sample size

Note that the finite correction factor is a number between zero and one. At the extreme, when a census is performed the sample size equals the population size, or n = N, the finite correction factor equals zero. At the other extreme when the population is infinite, N is much greater than n, and the finite correction factor approaches a value of one. The finite correction factor is multiplied to the measure of dispersion in estimating the confidence interval. The resultant equation for confidence interval when the population standard deviation, σ_x, is known and the sample is large is shown here:

$$\text{Confidence Interval} = \overline{X} \pm z \text{ score } \frac{\sigma_x}{\sqrt{n}} \sqrt{\frac{(N-n)}{(N-1)}}$$

where $\overline{X}$ is the sample mean

z-score is the value from the standard normal distribution corresponding to the confidence level

σ_x is the population standard deviation

n is the sample size

N is the size of the population

The steps of constructing a confidence interval once sample results have been analyzed can be summarized as follows:

1. Specify the level of confidence required on the interval.

2. Based on population parameters and sampling strategy, select the appropriate distribution

3. Use appropriate table or function to determine the critical value corresponding to the required level of confidence.

4. Compute the sample mean.

5. Compute the measure of dispersion based on whether known for the population or estimated from the sample.

6. Compute the confidence interval using the generic formulation provided.

This process can be illustrated with the aid of a simple numerical example. Consider that an auditor is estimating the average cost of inventory items. Suppose there are a million items in the inventory and the standard deviation of the population of inventory items is known to be $50. Follow the above step-by-step approach to construct the confidence interval:

- First, the auditor has to randomly select inventory items to value. In this example, 250 items of inventory will be selected and the costs of these items verified.

- Second, a confidence level has to be specified. The auditor wants to develop a 95% confidence interval. This means 95% of a similarly constructed confidence interval will contain the mean of 1 million inventory items.

- Third, the interval coefficient is appropriately determined. The sample size of 250 items is considered large, thus the standard normal distribution is used. A 95% confidence level translates to a z-score of 1.96.

- Fourth, the costs of the sample items are ascertained and the sample mean computed. Assume the sample mean is $110.

- Fifth, the measure of dispersion is computed. Because the population standard deviation is known, you can use it to determine the dispersion of the mean by dividing it by the square root of the sample size.

- Finally, the confidence interval can be estimated by using the following formula:

$$\overline{X} \pm z \frac{\sigma}{\sqrt{n}}$$

$$110 \pm 1.96 \frac{50}{\sqrt{250}}$$

or, 110 ± 6.20

Confidence Interval (103.80, 116.20)

Thus based on the sample results and the population standard deviation, with 95% confidence the auditor can assert that the true average value of a million inventory items is between $103.80 and $116.20.

In the previous example, instead of a million items of inventory, suppose there were only 1,000 items of inventory whose average value had to be estimated. Using the sample size of 250 items as before and the values of population standard deviation and sample mean as before, the resultant confidence interval can be computed by this equation:

$$\text{Confidence Interval} = \overline{X} \pm z \text{ score} \frac{\sigma_{\overline{x}}}{\sqrt{n}} \sqrt{\frac{(N-n)}{(N-1)}}$$

$$110 \pm 1.96 \frac{50}{\sqrt{250}} \sqrt{\frac{(1000-250)}{(1000-1)}}$$

$$110 \pm 1.96 \frac{50}{\sqrt{250}} \sqrt{\frac{750}{999}}$$

or, 110 ± 5.37

Confidence Interval (104.62, 115.37)

The reduction in the width of the confidence interval is attributed to the finite correction factor. As 25% of the items in the population were in the sample, the values of these items were ascertained by the auditor and are not subject to estimation. In other words, even though there are 1,000 items in the population, the average value of 750 of those items have to be estimated given the average values of the 250 items in the sample have been determined with certainty. This results in narrowing of the confidence interval.

The precision of the estimate is analogous to the width of the confidence interval. Higher degree of precision in estimating the population mean is obtained when the width of the confidence interval is narrow. As the width of the confidence interval increases, the precision of the estimation decreases. Thus, to increase the precision of the estimate the confidence interval has to be narrowed. One approach to increasing precision is to reduce the confidence level. As the confidence level decreases, the corresponding z-score reduces and the interval becomes narrower. However, there is a trade-off. Increase in precision is obtained at the expense of reducing confidence level. In other words, obtaining a narrow, more precise interval through reduction of confidence level increases the risk that the narrower interval might not contain the population mean.

A second alternative for increasing precision of an interval estimate is to increase the sample size. An increase in sample size reduces the measure of dispersion. Thus, the increase in sample size results in a reduction of the width of the confidence interval and thus increases precision. For a finite population, an increase in sample size also results in a higher proportion of the population being sampled. This increase in the proportion of the population being sampled reduces the finite correction factor, resulting in a further narrowing of the interval and consequently results in increased precision.

10.7 Confidence Interval for a Large Sample When Population Standard Deviation Is Unknown

In many accounting and auditing applications, the population standard deviation is not known and has to be estimated from the sample data. Thus the sample provides an estimate of not only the population mean but also of the standard deviation. This is in contrast to the previous discussion in which the population mean was not estimated from the sample; instead, the standard deviation of the population was assumed to be known. The sample standard deviation denoted as s_x can be calculated using this formula:

$$s_x = \sqrt{\frac{\sum (X - \bar{X})^2}{(n-1)}}$$

where s_x is the sample standard deviation
$\bar{X}$ is the sample mean
n is the sample size

Large sample sizes enable the usage of z-scores as interval coefficient, based again on the specified level of confidence. Further if the sample size is large relative to the population, the finite correction factor should be used. The confidence interval equation for estimating population mean using a large sample size to estimate both the mean and the standard deviation is given as

$$\text{Confidence Interval} = \bar{X} \pm z \text{ score} \frac{s_x}{\sqrt{n}} \sqrt{\frac{(N-n)}{(N-1)}}$$

where $\bar{X}$ is the sample mean

z-score is the value from the standard normal distribution corresponding to the confidence level

s_x is the estimated sample standard deviation

n is the sample size

N is the size of the population

For an extremely large or infinite population size, the finite correction factor approaches a value of one and can be eliminated from consideration.

To demonstrate the use of the equation, look at this simple numerical example. Suppose a sample of 40 items is selected as given in Table 10.2. A 90% confidence interval of the sample mean is computed, considering two scenarios as the population is infinite; the population consists of 250 items.

Table 10.2 A Population of 40 Items for Illustration

Column 1	Column 2	Column 3	Column 4	Column 5	Column 6	Column 7	Column 8
30	22	18	45	37	26	42	12
25	36	33	47	44	26	20	34
28	46	15	41	42	29	16	37
31	30	48	28	40	22	18	42
21	17	29	36	33	47	25	31

Mean = = 31.225
Variance = σ^2 = 102.076
Standard Deviation = σ = 10.1

Recall that the mean of these 40 items was 31.225, and the standard deviation was 10.1. A 90% confidence level translates to a z-score of 1.645. Estimate the confidence interval for infinite population by using this formula:

$$\overline{X} \pm z \frac{s_x}{\sqrt{n}}$$

$$31.225 \pm 1.645 \frac{10.1}{\sqrt{40}}$$

or, 31.225 ± 2.627

Confidence Interval = (28.598, 33.852)

When the population size is small and a significant proportion of items are selected as the sample, a finite correction factor is multiplied to the measure of dispersion. The resultant confidence interval, for this example when the population size is 250, can be computed as

$$\text{Confidence Interval} = \overline{X} \pm z \text{ score } \frac{s_x}{\sqrt{n}} \sqrt{\frac{(N-n)}{(N-1)}}$$

$$31.225 \pm 1.645 \frac{10.1}{\sqrt{40}} \sqrt{\frac{(250-40)}{(250-1)}}$$

$$31.225 \pm 1.645 \frac{10.1}{\sqrt{40}} \sqrt{\frac{210}{249}}$$

or, 31.225 ± 2.412

Confidence Interval = (28.812, 33.637).

As in the previous example, the precision of the estimate can be enhanced by reducing the range of the confidence interval. Recall, the range is defined as the difference between the upper bound and the lower bound of the interval. The range of the interval is based on three factors: the level of confidence, the standard deviation of the population or sample and the sample size. Reducing the level of confidence in the interval narrows the interval and increases the precision of the estimate. However, this may not be desirable as the risk that the interval does not contain the population mean is higher. The standard deviation could be reduced, but that also requires increasing sample size when the standard deviation is estimated from the sample. Hence the only recourse available to accountants to enhance the precision is through increasing the sample size. An increase in sample size reduces the standard error of the sampling distribution, thereby reducing the range. It should be noted however, that increase in sample size has a diminishing marginal return on reducing the range or width of the confidence interval.

10.8 Confidence Intervals for Small Samples

In an accounting context, at times, sampling more than 30 items may not be possible or prohibitively expensive. Consider an auditor assessing a client's acquisition of other businesses. Suppose the client had acquired 75 smaller businesses in the current accounting period and the auditor is determining whether the fair value of the acquisition was appropriately determined. As it is time-consuming and expensive to perform a detailed audit of each acquisition to determine fair values of each asset and liability account, the auditor may have to limit his sample to 7 or 8 acquisitions. When the sample size is less than 30, the requirements of applying normal distribution are not met. Rather, the auditor has to use other statistical techniques that are appropriate when dealing with a small sample size.

When the sample size is small and statistical inferences have to be made on the population characteristic, an assumption of normality for the population distribution is critical.

For small sample sizes drawn from a population, which is normally distributed, the sample mean is known to be distributed as a t-distribution. This distribution was proposed in 1908 by a brewmaster in Ireland, W.S. Gosset, who published his finding under a pseudo name of Student. Hence, the distribution is often referred to as the Student-t distribution. Formally stated, when the population is normally distributed with an unknown standard deviation, the sampling distribution of a small sample of n can be described by a t-distribution with (n-1) degrees of freedom.

Two properties of t-distribution are noteworthy. First, the distribution is symmetrical and ranges from -∞ to +∞. Second, the standardized t-distribution has a mean of zero and a standard deviation equal to

$$\sqrt{\frac{n-1}{(n-1)-2}}\,\sigma \text{ for } n \geq 4$$

These two properties indicate the similarities between the t-distribution and a standardized normal distribution and also point out one important difference. Both the t and Z distributions are symmetrical and range from −∞ to +∞. Both have a mean of zero. The difference, however, lies in the spread of the two distributions. The variance and standard deviation of a z-distribution is equal to one, whereas the variance and standard deviation of a t-distribution is greater than one. In fact, the standard deviation of a t-distribution depends on the sample size. As the sample size increases, the standard deviation decreases. A plot of t-distribution for various sample size is shown in Figure 10.6. The plots in the figure illustrate the change in the variance of the distribution as the sample size increases. As sample size approaches 30, the variance approaches 1, and the plot is approximately the same as that of the standardized normal.

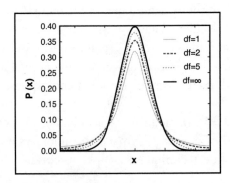

The distribution for df = ∞ is analogous to a normal distribution.

Figure 10.6 Probability distribution function of t-distributions with varying degrees of freedom

As with the standardized normal distribution, tables of t values have also been developed. These are also available on a scientific calculator and on Excel. The t-value for a 90% confidence interval with sample size 10, for example, is 1.833. This compares to the 1.645 for the normal distribution. Likewise, the t-value corresponding to a 95% confidence level for sample size of 14 is 2.160 and that for 8 is 2.306. This compares with 1.96 for the z-score of a standardized normal.

Using the t-statistics in place of the z-score makes the confidence interval wider. This, as expected, reduces the precision of the estimate. As in the previous sections, the width of the confidence interval can be reduced, or precision enhanced, by increasing the sample size. For a t-distribution increasing sample size helps in two ways. First, it reduces the measure of dispersion. Second, it reduces the corresponding t-statistics by increasing the degrees of freedom.

The steps to numerical computations are equivalent to the ones just shown. The confidence interval can be similarly obtained by substituting the t-statistics in place of z-score.

10.9 Confidence Intervals for Proportions

The methodology of confidence intervals can be extended to discrete probability distributions. The estimation of confidence intervals for a binomial distribution is discussed next. This is of relevance to forensic accountants when determining an error rate for a population. In attribute sampling, such as in tests of control, the greater concern to the auditor is the error proportion in the sample, denoted as $\hat{p}$. This proportion is determined as

$$\hat{p} = \frac{X}{n}$$

where $\hat{p}$ = error proportion in sample

X is the number of errors found in the sample

n is the sample size

Recall the sampling distribution of $\hat{p}$ values can be approximated by a normal distribution if the sample size is large and the population proportion is not too close to zero or one. The confidence interval of the population proportion based on the knowledge of sample proportion $\hat{p}$ is constructed as

$$\text{Confidence Interval of Population Proportion} = \hat{p} \pm z \text{ score} \sqrt{\frac{\hat{p}(1-\hat{p})}{n}}.$$

where $\hat{p}$ is the error proportion in the sample

Z is the interval coefficient

n is the sample size

To illustrate this equation take a look at this simple numerical example. Consider that the auditor is interested in finding the error rate in an internal control procedure. Suppose the supervisor has to sign the shipping document prior to the goods being shipped. The auditor randomly selects 300 documents and finds that 54 documents do not have the supervisor's signature. A 95% confidence interval can be constructed as

$$\hat{p} = \frac{54}{300} = 0.18$$

$$\text{Confidence Interval} = 0.18 \pm 1.96 \sqrt{\frac{0.18(1-0.18)}{300}}$$

Confidence Interval = 0.18 ± 0.0435

Confidence Interval = (0.1365, 0.2235)

Thus, there is a 95% chance that the interval (0.1365, 0.2235) contains the population proportion of missing supervisor signature in shipping documents.

10.10 Chapter Summary

In forensic accounting and auditing, the auditor relies on sampling procedures to collect pertinent audit evidence. Due to time and cost constraints a complete examination and verification of all accounts or transactions is not feasible. Hence the auditor often relies on sampling procedures to obtain necessary evidence. Chapter 9, "Sampling Theory and Techniques," covered the proper sampling procedures to be undertaken so that the sample results can be generalized to the population at large. This chapter presented the mathematical techniques through which such generalizations are operationalized.

The use of sampling to obtain evidence on the population and the consequent use of confidence intervals is important in a forensic accounting and auditing context. The following list summarizes the necessary steps to consider when constructing a confidence interval.

1. Define the attribute to be tested.

2. Define the population on which the results will be generalized with a high degree of specificity. The population cannot be modified to include or exclude items once a sample has been selected. If the population changes at the time of conducting the test, the entire sample and results have to be abandoned, and a new sample has to be selected.

3. Assess a-priori the information needs of the user and the desired level of confidence in the results as well the desired level of precision of the result.

4. Determine the requisite sample size that would likely provide the results at the desired level of precision. This topic is discussed in the next chapter.

5. Use the probabilistic sampling method to select a random sample if the sample results are to be generalized to the population. Note that use of nonprobabilistic sampling methods, such as judgment sampling, though may be useful in certain situations, cannot be generalized to the population.

6. Based on the sample size and assumption regarding the underlying probability distribution, determine the appropriate test statistic to use. Figure 10.7 shows a flowchart to determine the appropriate test statistic for different sample sizes and assumptions. Use tables, Excel, or a scientific calculator to convert the level of confidence specified in Step 3 to the corresponding value of the test statistic.

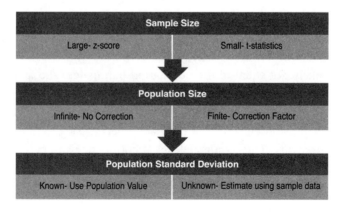

Figure 10.7 Flowchart for confidence interval estimation

7. Compute the sample statistics, which include the mean and standard deviation, in the sample.

8. When the population standard deviation is known, it is used in constructing the confidence interval. In situations when the population standard deviation is unknown, it is estimated from the sample.

9. Use a finite correction factor when the sample size is greater than 5% of the population.

10. Compute the confidence interval using the appropriate equation. The use of equations is outlined in Table 10.3.

Table 10.3 Relevant Parameters to Estimate Confidence Intervals

Sample Size	Estimation of Dispersion	Population Size	Distribution to be Used	Measure of Dispersion	Finite Correction Factor
Large	Population Standard Deviation Known	Infinite	z-Score	Standard Deviation of the Population	None Required
Large	Population Standard Deviation Known	Finite	z-score	Standard Deviation of the Population	$\sqrt{\frac{(N-n)}{(N-1)}}$
Large	Population Standard Deviation Unknown	Infinite	z-score	Sample Standard Deviation	None Required
Large	Population Standard Deviation Unknown	Finite	z-score	Sample Standard Deviation	$\sqrt{\frac{(N-n)}{(N-1)}}$
Small	Population Standard Deviation Unknown	Finite	t-statistics	Sample Standard Deviation	None Required

Although the use of point estimates is easy, it is of limited usefulness as the value of the point estimate is likely not equal to the population parameter. The quality of the point estimator is enhanced when it is unbiased, consistent, efficient, and sufficient. Even with a high quality point estimator, the need for another measure that provides not just the estimate but a level of confidence on that estimate is necessary. Confidence interval estimates fulfill that necessity. The estimate's level of confidence and precision can be controlled through a trade-off between these factors.

An important but subtle point regarding the interpretation of confidence interval is that the probability judgment is on whether the interval consists of the population mean and not that the population mean is between the lower and upper bound of the interval. It is important to remember that the population mean, though unknown, is a fixed value; the randomness arises due to selection of the sample and hence in determining the confidence interval. The confidence interval either contains the population mean, or it does not. Thus, the term 95% confidence interval implies that 95% of the intervals similarly constructed will contain the population mean. There is no way of assessing whether the particular confidence interval includes the population mean. However, for a 95% confidence interval, only 5% of a similarly constructed interval will not include the population mean—others will. If the confidence level is increased and the interval widened, it increases the probability that the population mean is contained in the wider interval. Increasing the sample size, if possible, also provides a greater degree of comfort on the sample results. The following chapter discusses the determination of adequate sample size.

Appendix: Use of a Scientific Calculator and Excel to Compute Confidence Intervals

Using the Scientific Calculator to Compute Confidence Interval

Set up Procedure:

- Press STAT.
- Arrow Right to TEST.
- Select 7:ZInterval and press Enter.

Choose either DATA or STATS.

In DATA the calculator will read from a LIST.

In Stats mode the setting can be entered for standard deviation, mean, sample size, and the level of confidence.

View Display and highlight "Calculate" and press Enter.

The result will be displayed as (lower bound, upper bound).

Using Excel to Compute Confidence Interval

Use the formula function CONFIDENCE(

Within the parentheses, enter

- alpha, or (1 − level of confidence), that is, for a 95% confidence interval, type in 0.05;
- standard deviation; and
- sample size.

Close parentheses and press "enter."

The number returned is one-half the width of the confidence interval. Add and subtract this number from the mean to obtain the desired confidence interval.

10.11 Endnote

1. Systematically ignoring such rare events leads to errors which could be momentous in financial markets as elaborated in the *The Black Swan* by Nassim Nicholas Taleb.

11

Determining Sample Size

Those who do not believe in sampling should never go for a blood test.

11.1 Introduction

In a statistical context, a "sample" of the population can yield a reliable estimate of population statistics provided that (i) each item in the population has an equal, unbiased chance of being selected, (ii) that the sample is properly stratified to take into consideration the characteristics and risk factors of the population, and (iii) the sample is large enough to be representative of the characteristics of the population. Accordingly, a sampling strategy has to be designed to facilitate the statistical projection of errors.

The process of determining appropriate sample size entails specifying a level of reliability and a degree of precision. In the statistical context, "reliability" is the level of assurance or confidence (expressed as a percentage) that the statistical results will provide reliable information about the population as a whole. Precision is determined in statistical sampling by estimating how much error is expected to be found and how much error can be tolerated. After the sample is selected and results obtained, there is a trade-off between reliability and precision. The reliability of the results can be increased by reducing precision and vice versa. Prior to selecting the sample, however, the auditor could achieve both higher precision and higher reliability by increasing the sample size. Hence, the purpose of obtaining statistical results and the corresponding information needs have to be assessed prior to selecting the sample.

This chapter presents the statistical methodology of selecting an appropriate sample size for various situations. First discussed are situations when the standard deviation of the population is known. However, in most real-world cases the standard deviation has to be estimated from sample (approximations to estimate sample size are given). Finally, the chapter covers the determination of sample size for proportion. You'll find two tables that provide sample sizes for common requirements of precision and level of confidence.

11.2 Computing Sample Size When Population Deviation Is Known

In an accounting and auditing context, the auditor usually has a desired confidence level and a specified precision level but is unsure about what sample size will attain those objectives. Because testing and verifying the sample items is usually costly, the auditor is sensitive to the number of items that have to be audited. The auditor wishes to optimize on the sample size such that it is the minimum required to achieve the audit objectives. This section discusses the determination of sample size when the standard deviation of the population is known.

To start, let's introduce a new measure, the tolerable error. The term *tolerable error* has been discussed extensively in audit sampling guidance. In the context of statistics and the formulation in this book, tolerable error is defined as one-half the width of the confidence interval. Recall from the previous chapter that

$$\overline{X} \pm z \frac{\sigma}{\sqrt{n}}$$

Thus the width of the confidence interval is $2z \frac{\sigma}{\sqrt{n}}$, and the tolerable error, ε is

$$\varepsilon = z \frac{\sigma}{\sqrt{n}}$$

where, ε is the tolerable error, z is the z-score of a standardized normal corresponding to the desired level of confidence, σ is the population standard deviation which is known, and n is the sample size.

Now consider a numerical example. Suppose the auditor is assessing the average weekly amount paid to day laborers and knows that the population standard deviation is $50. Further, suppose the auditor desires a 90% confidence and a precision of $16 or a maximum width of confidence interval to be $16. As tolerable error, ε, is half the confidence interval, the auditor is willing to accept a tolerable error of $8. The values can be substituted in the given formula and solved for the sample size, n, as follows

$$\varepsilon = z \frac{\sigma}{\sqrt{n}}$$

Rearranging terms you get

$$n = \frac{z^2 \sigma^2}{\varepsilon^2}$$

Substituting z = 1.645 corresponding to a 90% confidence, σ equal to 50 and ε equal to 8, you get

$$\frac{(1.645)^2 (50)^2}{8^2}$$

or

n = 105.70

Hence, rounding up, select a sample of 106 persons to obtain the desired level of confidence for a tolerable error of $8.

Increasing the desired level of confidence will increase the sample size. Also, decreasing the tolerable error, thereby increasing the level of precision, will also increase the sample size. Suppose in the example just given the auditor desires a 95% confidence for a precision of $10. Thus, the desired width of the confidence interval is $10, which implies that the tolerable error is $5. The confidence level of 95% converts to a z-score of 1.96. Substituting these values into the equation, the sample size is

$$n = \frac{z^2 \sigma^2}{\varepsilon^2}$$

Substituting z = 1.96 corresponding to a 95% confidence, σ equal to 50 and ε equal to $5, you get

$$n = \frac{(1.96)^2 (50)^2}{5^2}$$

or

n = 384.16

Hence, rounding up, select a sample of 385 items to obtain the desired 95% confidence for a tolerable error of $5. The auditor may or may not be able to justify additional costs of testing and verifying additional items. In that case, he might not afford increasing the sample size from 106 to 385 to obtain increased confidence and precision. In practical situations, auditors are often forced to strike a balance between required confidence and precision and the sample size they can afford.

11.3 Sample Size Estimation when Population Deviation Is Unknown

In the previous section, the population standard deviation was assumed to have been known. This type of situation can exist in an audit context for long-time audit clients of an audit firm. For example, Deloitte has audited General Motors for many years, and the past years' work-papers and experience can enable Deloitte to assess the population standard deviation. The assessment is reasonable if the past audit experience on estimation were not erratic or excessively unstable.

In many cases an estimate of the population standard deviation might not be available from past experience. It might be possible, however, to get a rough estimate of the standard deviation if there is some knowledge of the total range of the basic random variable. The range is defined as the difference between the largest and smallest value of the population. A good approximation of standard deviation is one-sixth of the range. This is based on the reasonable assumption that the distribution is spread by three standard deviations from the mean. Recall that for a normal distribution 99.75% of observations lie within three standard deviations from the mean. That is,

$$P(\mu - 3\sigma \leq x \leq \mu + 3\sigma) = 0.9975$$

In situations when the population is non-normal, this estimation may not be exact. Nevertheless, use of this reasonably good approximation of the standard deviation yields a workable sample size. This is especially applicable in situations where there is no basis for estimating the standard deviation. One of the major advantages of this method of a-priori estimating the deviation is its simplicity and ease of use. This is a three-step process:

1. Sort the database and identify the maximum and minimum balances.
2. Compute the difference between the maximum and minimum balance.
3. Divide the difference by six.

This yields a workable approximation of the standard deviation for the purpose of computing the required sample size.

Let's continue with the previous example of verifying the average weekly amount paid to contract laborers, but now the auditor does not know the standard deviation in the population. Instead he knows that the maximum allowed under company policy is $600, and the minimum allowed by law is $200. Although the auditor does not know if any contract laborer was in fact paid either of those amounts, the upper and lower bound provide a conservative and defensible basis to compute required sample size. Continuing with a desired confidence of 90% with a desired precision of $16, the required sample size can be computed following these steps:

1. 90% confidence level translates to a z-score of 1.645
2. Range of $400 converts to a conservative estimate of σ as 66.67
3. Desired precision of $16 translates to a tolerable error of $8
4. Substituting these values in the equation to compute n, we get

$$n = \frac{(1.645)^2 (66.67)^2}{8^2} = 187.93$$

Rounding up, the required sample size is 188 people.

Table 11.1 provides sample size for differing confidence levels and differing degrees of precision. The degree of precision is denoted as a percentage of the range of values of the variable. When the standard deviation of the population is unknown, it can be estimated by dividing the range by six. Thus, both standard deviation and the degree of precision are presented in terms of the range. Table 11.1 presents sample sizes for three levels of confidence, 90%, 95%, and 99%. For each level of confidence, 8 degrees of precision is presented. The degree of precision is given a value between 2% and 10% of the range. In total, samples size for 24 scenarios arising from 3 levels of confidence and 8 levels of precision are computed in Table 11.1.

Table 11.1 Sample Sizes for Differing Levels of Confidence and Precision for a Continuous Variable

Row #	Level of Confidence	Z	Precision as a Percentage of Range	Sample Size
1.	90%	1.645	10%	30
2.	90%	1.645	8%	47
3.	90%	1.645	6%	84
4.	90%	1.645	5%	121
5.	90%	1.645	4%	188
6.	90%	1.645	3%	335
7.	90%	1.645	2.5%	482
8.	90%	1.645	2%	752
9.	95%	1.96	10%	43
10.	95%	1.96	8%	67
11.	95%	1.96	6%	119

Row #	Level of Confidence	Z	Precision as a Percentage of Range	Sample Size
12.	95%	1.96	5%	171
13.	95%	1.96	4%	267
14.	95%	1.96	3%	475
15.	95%	1.96	2.5%	683
16.	95%	1.96	2%	1,068
17.	99%	2.576	10%	74
18.	99%	2.576	8%	116
19.	99%	2.576	6%	205
20.	99%	2.576	5%	295
21.	99%	2.576	4%	461
22.	99%	2.576	3%	820
23.	99%	2.576	2.5%	1,180
24.	99%	2.576	2%	1,844

By reviewing Table 11.1 you can see that the sample size increases as the level of confidence increases for the same degree of precision. This can be observed by comparing Rows 4, 12, and 20 or Rows 6, 14, and 22, and so on. Additionally, by comparing any set of consecutive rows, 1–8, 9–16 or 17–24, it can be observed that for the same level of confidence the sample size increases as the degree of precision increases, as measured by a smaller percentage of the range.

Additionally, by comparing the sample sizes in Table 11.1 for different scenarios, you can see that the level of precision has the larger impact on the sample size. For example, comparing Row 4 to Row 12 and 20, one can evaluate the effect of increasing the level of confidence while keeping the degree of precision the same. That is, for precision equal to 5% of the range, the increase in the level of confidence from 90% to 95% to 99% increases the sample size from 121 to 171 to 295. That is an increase of sample size of about one and a half times to increase the confidence level from 90% to 99%. In contrast, increasing the degree of precision yields a much sharper increase in sample size. This can be observed by comparing Row 4 to Row 7. The level of confidence in both these scenarios equals 90%. However the degree of precision is reduced from 5% to 2.5%. The resultant sample size increases from 124 to 482 or a two and a half-fold increase. Thus, increasing the level of precision has a greater impact on sample size than increasing the level of confidence while maintaining the same level of precision.

On occasions, this formula may yield a small sample size that is sufficient to attain the desired level of confidence and precision. Unfortunately, the assumptions and the formula for small samples differ from those for large samples. Therefore, if the formulas result in a sample size of less than 30, the simplest and safest strategy is to select a sample of 30 items. The alternative would be to abandon normality assumptions and recompute the sample size for an unknown distribution, which often could yield a sample size greater than 30. Determining sample size is not an exact science, and neither is it critical. Selection of a larger than optimal sample is marginally inefficient but does not compromise the effectiveness of the statistical results. On the contrary, a sample size that is greater than the minimum might yield a higher than desired level of precision at the conclusion of the statistical procedure.

11.4 Sample Size Estimation for Proportions

Likewise, when estimating proportions or rates in a binomial sample, the level of confidence and the level of precision could be specified. The desired level of confidence and the level of precision can be used to determine the requisite sample size. Again, the tolerable error is defined as half of the level of precision, the width of the confidence interval.

The requisite sample size to estimate a proportion with a specified tolerable error and a desired level of confidence can be obtained through the formulation of tolerable error. Recall from the previous chapter that the confidence interval for a proportion is given by the following equation

$$\mu \pm Z\sqrt{\frac{p(1-p)}{n}}$$

The width of confidence interval or the precision is therefore equal to

$$2Z\sqrt{\frac{p(1-p)}{n}}$$

and the tolerable error, $\varepsilon = Z\sqrt{\frac{p(1-p)}{n}}$

Rearranging terms we get

$$n = \frac{Z^2}{\varepsilon^2}p(1-p)$$

When the value of p is unknown, the measure of standard deviation, $\sqrt{p(1-p)}$, cannot be computed. The value of p can be estimated from prior experience. In situations when there is no prior experience, the value of p could be conservatively chosen as 0.5. A value of 0.5 for p yields the maximum value of [p(1-p)] as 0.25. Every other value of p would result in p(1-p) being less than 0.25. In statistics, this is called the maximum entropy principle. Hence, assumption of p to be 0.5 will yield the most conservative sample size.

This formula for determining sample size can be illustrated with a simple numerical example. Consider that the auditor is assessing the error rate in the valuation of inventory items. The desired level of confidence is 0.95, and the desired level of precision is 0.04, which is a confidence interval that is ±0.02 around the mean. Say the auditor expects that the highest possible error rate in valuation is 0.1. Substituting anticipated maximum error rate $p = 0.1$, tolerable error $\varepsilon = 0.02$ and $Z = 1.96$ corresponding to a confidence level of 95% in the equation

$$n = \frac{Z^2}{\varepsilon^2} p(1-p)$$

we get

$$n = \frac{(1.96)^2}{(0.02)^2}(0.1)(1-0.1)$$

$$n = 864.36$$

Rounding up, a verification of a sample of 835 inventory items will provide a 95% confidence within a 0.04 level of precision of the true error rate in valuation of the inventory population. Table 11.2 provides sample size for differing confidence levels, degree of precision, and anticipated error rate. Table 11.2 presents sample sizes for three levels of confidence, 90%, 95%, and 99%. Sample size is computed for three levels of the degree of precision for each level of confidence, 0.1, 0.05, and 0.02. Finally, for each level of confidence and precision, the sample size is given for four levels of anticipated error rate, 0.05, 0.1, 0.25, and 0.5. In total, samples size for 36 scenarios arising from three levels of confidence, three levels of precision, and four anticipated error rates, are computed in Table 11.2.

Table 11.2 Sample Sizes for Differing Levels of Confidence, Precision, and Anticipated Error Rate for Proportions

#	Confidence Level	Z	Precision Level	ε	Anticipated Error Rate (p)	Entropy p(1-p)	Sample Size
1.	90%	1.645	0.1	0.05	0.05	0.0475	52
2.	90%	1.645	0.1	0.05	0.1	0.09	98
3.	90%	1.645	0.1	0.05	0.25	0.1875	203

#	Confidence Level	Z	Precision Level	ε	Anticipated Error Rate (p)	Entropy p(1-p)	Sample Size
4.	90%	1.645	0.1	0.05	0.5	0.25	271
5.	90%	1.645	0.05	0.025	0.05	0.0475	206
6.	90%	1.645	0.05	0.025	0.1	0.09	390
7.	90%	1.645	0.05	0.025	0.25	0.1875	812
8.	90%	1.645	0.05	0.025	0.5	0.25	1,083
9.	90%	1.645	0.02	0.01	0.05	0.0475	1,286
10.	90%	1.645	0.02	0.01	0.1	0.09	2,436
11.	90%	1.645	0.02	0.01	0.25	0.1875	5,074
12.	90%	1.645	0.02	0.01	0.5	0.25	6,766
13.	95%	1.96	0.1	0.05	0.05	0.0475	73
14.	95%	1.96	0.1	0.05	0.1	0.09	139
15.	95%	1.96	0.1	0.05	0.25	0.1875	289
16.	95%	1.96	0.1	0.05	0.5	0.25	385
17.	95%	1.96	0.05	0.025	0.05	0.0475	292
18.	95%	1.96	0.05	0.025	0.1	0.09	554
19.	95%	1.96	0.05	0.025	0.25	0.1875	1,153
20.	95%	1.96	0.05	0.025	0.5	0.25	1,537
21.	95%	1.96	0.02	0.01	0.05	0.0475	1,825
22.	95%	1.96	0.02	0.01	0.1	0.09	3,458
23.	95%	1.96	0.02	0.01	0.25	0.1875	7,203
24.	95%	1.96	0.02	0.01	0.5	0.25	9,604
25.	99%	2.576	0.1	0.05	0.05	0.0475	127
26.	99%	2.576	0.1	0.05	0.1	0.09	239
27.	99%	2.576	0.1	0.05	0.25	0.1875	498
28.	99%	2.576	0.1	0.05	0.5	0.25	664
29.	99%	2.576	0.05	0.025	0.05	0.0475	505
30.	99%	2.576	0.05	0.025	0.1	0.09	956
31.	99%	2.576	0.05	0.025	0.25	0.1875	1,991
32.	99%	2.576	0.05	0.025	0.5	0.25	2,655
33.	99%	2.576	0.02	0.01	0.05	0.0475	3,152
34.	99%	2.576	0.02	0.01	0.1	0.09	5,973
35.	99%	2.576	0.02	0.01	0.25	0.1875	12,443
36.	99%	2.576	0.02	0.01	0.5	0.25	16,590

By reviewing Table 11.2 it can be observed that the sample size increases as the level of confidence increases for the same level of tolerable error and anticipated error rate. This can be seen by comparing Rows 1, 13, and 25, or Rows, 2, 14, and 26 etc. Further, by comparing Rows 1, 5, and 9 or 14, 18, and 22, etc., it can be observed that for the same confidence level and anticipated error rate, sample size increases as the tolerable error decreases. Additionally, by comparing any four consecutive rows, it can be observed that for the same level of confidence and precision the sample size increases as anticipated error rate increases to 0.5.

Additionally, by comparing the sample sizes in Table 11.2 for different scenarios, it can be observed that the level of precision has the largest impact on the sample size. For example, comparing row number 2 to row number 8, you can see that reducing the level of precision from 0.1 to 0.02 increased the sample size from 98 to 2,436, approximately a 24-fold increase. On the other hand, increasing the level of confidence from 90% to 95% for a precision level of 0.05 and anticipated error rate of 0.1 increases the sample size from 390 to 554, only a 42% increase. This can be seen by comparing row number 6 to row number 18. Further, increasing the level of confidence to 99% for the same level of precision of 0.05 and anticipated error rate of 0.1 increases the sample size to 956, or a one and a half-fold increase. Thus, increasing the level of precision has a greater impact on sample size than increasing the level of confidence while maintaining the level of precision.

A caveat to using the methodology presented in this section is that the underlying assumption of this formulation is that either the sampling is conducted with replacement or that the sample size is less than 5% of the population size. If both of these conditions are violated, the analysis presented in this section is not applicable. In those cases a different probabilistic model has to be developed to guide inferences from the sample results.

11.5 Chapter Summary

In this chapter you learned about an important and useful practical tool for forensic accountants and auditors. Determination of adequate sample size for a variety of audit and forensic accounting tasks is a critical function of the accountant. This chapter presented techniques to determine sample sizes for different probabilistic assumptions, with or without a-priori knowledge of the deviation in the population. Approximations were developed for practical situations when there is a lack of knowledge on important population characteristics.

Two important characteristics that affect the sample size are the level of confidence and the degree of precision. The level of confidence is a measure that signifies the probability that the resultant confidence interval would include the population mean. The higher the level of confidence, the more likely the resultant interval is to contain the population mean. The degree of precision is a measure that signifies the resultant width of the confidence interval. For example it can be said with certainty that the mean of any population is between negative infinity and positive infinity, although the information is of no value. In various accounting and auditing applications, the task is to ascertain the level of confidence on a very narrow range of values, hence the degree of precision is high. There are of course trade-offs between the level of confidence and the degree of precision once the sample has been evaluated. However, at the outset, a requisite sample size can be determined by incorporating both the desired level of confidence and precision in the computation of the sample size.

You were also presented with two tables that present sample size for differing degrees of confidence and precision. The computations and results are generic and hence can be used in a variety of situations.

12

Regression and Correlation

Prediction is very difficult, especially if it's about the future.
—Niels Bohr, Physicist (1885–1962)

12.1 Introduction

Fortunately, in forensic accounting we are not in the business of predicting the future; rather, we unravel the past. Probability concepts were introduced in Chapter 6, "Transitioning to Evidence," by defining marginal, joint, and conditional probabilities. Subsequent chapters introduced various probability distributions to assess the marginal probability of a variable. Also you learned about techniques for estimating the parameters of variable based on sample information. These methods and techniques were for each variable, in isolation. This chapter discusses situations when the value of a variable is dependent on the value of another variable, when both are continuous variables. The notion of joint and conditional probabilities developed in Chapter 6 for binary or discrete variables still apply. However, those have to be expanded to the continuous domain when both variables are continuous.

So what happens when the probability distribution of one variable is dependent on the value of another variable? Consider the distribution of housing prices in a town (earlier chapters treated this variable in isolation and developed mechanics of estimating the parameters of such variables through sampling and constructing confidence intervals). However, everyone knows that house prices are related to the square footage of the house. Both housing prices and square footage are variables, hence a joint probability distribution can be enumerated. Further, because the two variables are dependent, the conditional probability distribution of a home price, given the square footage, is not equal to the unconditional distribution. In this chapter the discussion is limited to the simplest of relationship between two variables, the straight line model. This methodology of estimating the value of one variable using a straight line relationship with another variable is called the *simple linear regression*. In general, however, the relationship between two variables could be curvi-linear or exponential. Statistical techniques have been developed to model such relationships but require logarithmic and other transformation to the data, which is beyond the scope of this book.

First in this chapter, the foundation for probabilistic relationship between two variables is built, followed by a discussion of correlation between two variables, which could either be positive, negative, or there can be no correlation between two variables. Next the fundamental criteria for determining the best fitting regression line and related measures are covered. The concept of p-value to test for statistical significance between two variables is also developed, followed by a discussion of the limitations and caveats of using regression analysis. The chapter concludes with a chapter summary and tips on performing regression analysis on Excel in the appendix.

12.2 Probabilistic Linear Models

Consider a linear model between two variables in physical sciences. The location of an object can be determined by its location at the start plus the product of speed "b" and time "x." This relationship is represented in graphical terms as

$y = \alpha + \beta x$

where, α is the y-intercept, or the point where the line intersects the y-axis,

β denotes the slope of the straight line, or the rate of change

x and y are the variables

This is a deterministic model because there is no provision for the value of y to vary once the value of x is known. The model implies that the variable y is known with certainty once x is known. In other words, there is no allowance for error in prediction. This model is applicable in many physical sciences and in some accounting relationships, such as the amount owed given the balance paid. However, in most accounting situations, the relationship is not deterministic, rather probabilistic. For example, although the amount of accounts receivable balance outstanding at the end of the period is related to the amount of total sales of the period, the relationship is not deterministic. Thus the relationship between the accounts receivable balance at the end of the period and the total sales for the period can be modeled through a probabilistic linear model. The probabilistic linear model introduced a component for random error, ε, in the equivalent deterministic model resulting in the equation

$y = \alpha + \beta x + \varepsilon.$

Note that the probabilistic model can be decomposed into two parts, the deterministic component and the random error. In such models, where the variable y is presented as a linear function of variable x, y is termed the *dependent variable* and x the *independent variable*.

12.3 Correlation

In studying the relationship between two variables of interest it is useful, at first, to plot the data on a graph. This enables visual examination of data to see whether the variables are related and if so, how. The initial evaluation of data allows the choice of appropriate statistical model to be developed and validated. Such a chart of simply plotting the observation is known as a scatter plot. Examples of scatter plots are shown in Figure 12.1. When the points on the plot tend to move from bottom left to top right, the two variables increase in tandem, and the relationship is termed as *direct*. In contrast, when the points on the plot move from top left to bottom right and an increase in one variable reduces the value of the other, the relationship is called *inverse*. If the relationship resembles a straight line, the relationship is linear. Instead if the points seem to fall on a curved line, the relationship is called a *curvilinear relationship*. Of course, the relationships are not always so straightforward. Sometimes the best line that fits the point is a horizontal line parallel to the x-axis. When that is the case, no relationship is said to exist between the two variables. The six panels of scatter plots in Figure 12.1 depict these relationships.

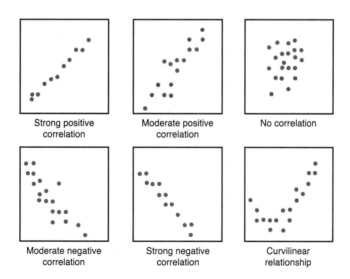

Figure 12.1 Scatter plots of data

Determining the degree of association or correlation between two variables is achieved through a correlation analysis. The metric termed coefficient of correlation quantifies the strength of relationship between two variables. The correlation coefficient takes a value between -1 and +1. Positive values of correlation coefficient denote a direct relationship, with the value of +1 depicting a perfectly direct relationship. Negative values of correlation coefficient denote an inverse relationship, with the value of -1 depicting a perfectly inverse relationship. The higher the absolute value of the correlation coefficient, the stronger the relationship. The correlation coefficient of zero denotes no relationship between the two variables.

12.4 Least Squares Regression

In a scatter plot of probabilistic linear relationship, as shown in various panels of Figure 12.1, multiple straight lines could be drawn through the data that would connect a few observations, and no line can be drawn that connects all the observations.

To determine which straight line best fits the data, statisticians have developed rules and metrics. First, a metric to measure error or deviation is needed. The error is measured as the vertical distance of each point from the straight line. When the point falls on the line the distance is zero. The data point could lie above or below the straight line, resulting in positive or negative values of deviation. These measures would cancel each other out when aggregated. Hence, another measure based on squaring the differences is developed. This is termed the *sum of the squared errors* or *SSE* because of squaring more weight is given to greater deviations from the line. To define the best fitting line, statisticians have developed common sense criteria. First, the line drawn has to be in the "center" of the data; that is, the sum of deviations from the line should equal zero. Second, the line should minimize the SSE or the sum of squared errors. The straight line that meets both the criteria for a data-set is called the *regression line*. The statistical method that uses this criteria is aptly termed Ordinary Least Squares Regression or *OLS regression*. Clearly, other measurement of errors and other criteria of optimizing the measures are possible, and those lead to alternate methods and regression lines and may be applicable in certain situations. However, the most commonly used method of regression is the OLS regression.

The best fitting line based on OLS regression satisfies the following two conditions.

1. The sum of errors or deviations of data points from the line, as measured by the vertical distance, is zero.
2. It is the straight line that minimizes the sum of squared errors, when error is measured as the vertical distance between the data-point and the line.

The line fitted by the ordinary least squares has the property that the total of all squared deviations is less than the corresponding total for any other straight line that could have been drawn on the same data.

The regression plot for a data-set is shown in Figure 12.2. The vertical distance for each observation is computed by taking the difference between the observed value of the dependent variable and its predicted value. The predicted value of the dependent variable is on the regression line corresponding to the value of x, the dependent variable. The regression line passes through the means of both variables, $\bar{X}$ and $\bar{Y}$. All least square regression lines must pass through, and it is this property that satisfies Condition #1 just listed, which is the sum of deviations above and below the line equal to zero.

The computational steps of regression are not discussed in this chapter. The requisite arithmetic is elaborate and the process is of no relevance to the interpretation of the

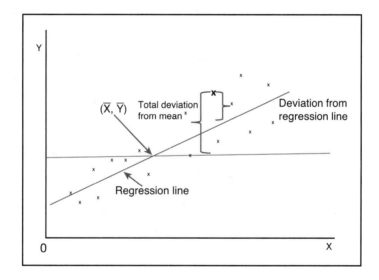

Figure 12.2   Regression plot for a data set

results. Instead the procedure for using either a scientific calculator or Excel to compute regression line is outlined in the appendix at the end of the chapter.

An important factor to remember when interpreting the results of regression is that the computations are based on a sample of data and not the entire population. Hence, there is a risk that the regression line fitted to the sample data would be different than the one that would fit the population. A measure known as the standard error, denoted as $s_{Y,X}$, estimates the standard deviation around the true but unknown regression line. This implies that to use the regression line to predict variables not included in the sample, you have to be cautious that the regression line that best fits the sample may be different from the one that fits the population due to sampling error.

12.5 Coefficient of Determination

The measure R^2 is known as the sample coefficient of determination for a regression equation. That is, this measure pertains only to the sample on which the regression parameters were estimated. The coefficient of determination is the percentage of the total variance in the dependent variable that is explained by the regression line. Figure 12.3 helps illustrate the situation. In Figure 12.3, one observation is considered. The total deviation of the observation Y from the mean is $(Y - \overline{Y})^2$. Once the regression line is fitted the vertical distance from the observation Y to the predicted value $\hat{Y}$ is the unexplained portion of the variance. The difference between the predicted value and the mean $(\hat{Y} - \overline{Y})$ is the part that is accounted for by the regression model. Algebraically, this can be represented as

Total Deviation = Deviation accounted for by Regression + Unaccounted Deviation

$$Y - \overline{Y} = (\hat{Y} - \overline{Y}) + (Y - \hat{Y})$$

Thus if the regression line rather than the mean value of the independent variable ($\overline{Y}$) was used to estimate the value of the observation (Y), the error would be reduced. There however is still an unexplained deviation, but the regression line has explained or accounted for a part of the deviation. The part of the deviation explained by the regression line divided by the total deviation is the value of the coefficient of determination. That is, an alternative interpretation of the coefficient of determination, R^2 is

$$R^2 = \frac{\text{Explained variation in Y}}{\text{Total variation in Y}}$$

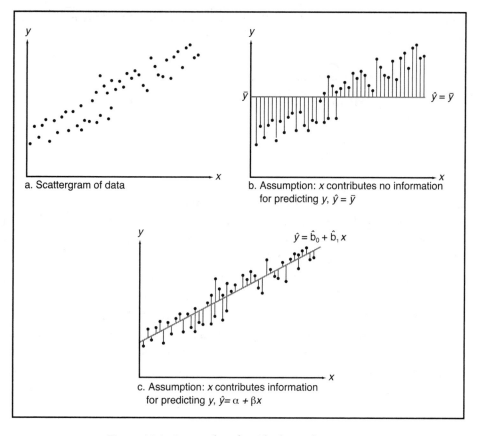

Figure 12.3 Scatterplot of residuals, ε, of regression

McClave, James T.; Benson, P. George; Sincich, Terry L.; *A First Course in Business Statistics*, 8th Ed., ©2001, p. 460. Reprinted and electronically reproduced by permission of Pearson Education, Inc., Upper Saddle River, New Jersey.

12.6 Test of Significance and p-Values

The null hypotheses, denoted as H_o, is a hypotheses that suggests that there is no relationship between the variables or the slope of the regression line equals zero. Complementary to the null hypotheses is the alternative hypotheses, denoted as H_1, that there is a relationship between the two variables or the slope of the regression line is nonzero. In testing the hypotheses you have to determine whether there is a relationship between the two variables. A slope of zero indicates no relationship. A non-zero slope, regardless of positive or negative, indicates a relationship between the two variables. The test-statistic is used to determine the significance of the result.

As with any test based on sample information, there is always a risk that the sample inference will not be true for the population. In terms of hypotheses testing this implies that the sample may accept or reject the hypotheses incorrectly. This is illustrated in Table 12.1.

Table 12.1   Description of Type I and Type II errors

	H_0 true in population	H_0 false in population
Sample accepts H_0	Correct	Type II error
Sample rejects H_0	Type I error	Correct

There are four possibilities between the two sample results and two population characteristics. The first situation is that the sample may accept the hypothesis, which in fact is true in the population. Hence a correct decision has been made by using sample information of accepting the null hypothesis. The second situation is that the sample may accept the null hypothesis when in fact it is false in the population. In that case an error has been made in accepting the hypothesis, which is not true. This type of error is termed as a Type II error and is denoted by β. The third situation is that the sample result rejects the hypothesis, which in fact is true in the population. A Type I error is made by rejecting a true hypothesis based on sample information and is denoted by α. The fourth situation is rejecting the null hypothesis based on sample information when it is false in the population. The probability of correctly rejecting a false hypothesis based on sample information is known as the power of the test and is measured as the complement of β or $(1 - \beta)$.

12.7   Prediction Using Regression

Two types of estimates can be made from regression. One is the estimate of conditional mean, which is given the value of the independent variable, the estimate of the average value of the dependent variable. The other is the estimate of an individual observation. The subtle difference between the two is not in the point estimate but the interval estimate. For the estimate of an individual observation the measure of dispersion is higher than for the estimate of the average. Single observations cannot be predicted with as much precision as the values of averages or means. The terminology commonly used to refer to the interval estimate of conditional mean is the confidence interval, and that for individual value is prediction interval. Clearly, the prediction interval is wider than the confidence interval.

A number of factors affect the width for both confidence intervals and prediction intervals. These are

- **Sample Size:** The larger the sample size, the smaller the standard errors and therefore narrower are the widths of the intervals.

- **Value of X for which prediction is being made:** The greater the deviation of the independent variable from its mean value, the greater the standard error and hence the width of the intervals. This means that the confidence intervals and prediction intervals are wider for very large and very small values of the dependent variables.

- **Standard error of the estimate:** The larger the standard error of the estimate, the greater the width of the intervals. That is, the greater the variability in the data, the less precise the prediction or estimation from the data.

- **Variability in the value of the independent variable:** The greater the variability in the value of the independent variable of the sample, the narrower the confidence interval and the prediction interval.

12.8 Caveats and Limitations of Regression Models

There are some key assumptions regarding the probability distribution of the random error, ε, in the OLS regression model. Recall from Section 12.2 that the probabilistic regression model contains two components, a deterministic model and a random error. The deterministic model, by definition, has no variability or uncertainty. Hence all the variability and uncertainty in a regression model arises from the uncertainty and variability in ε, the random error. Thus the assumptions on the probability distribution of ε are critical in evaluating the appropriateness of the regression model. These assumptions are

- The error term related to each observation is independently and identically distributed. This is known as *iid assumption*.

- The mean of the probability distribution of each ε is zero. That is, the average values of ε over a long series of observations would be zero. Hence, if each value of the independent variable is repeated multiple times and the average value of the dependent variable is plotted, the resultant model will be the deterministic model, as the error terms averaged over large number of observations would be zero for each and every value of the independent variable.

- The variance of the error term ε is constant and not dependent on the value of the independent variable. That is, higher values of the independent variable does not cause higher error terms.

- Finally, the error term is normally distributed.

Figure 12.4 pictorially illustrates the assumption on the probability distribution of the random error. The regression line is the solid line. The bell curves centered on the line denote the normal distribution for particular values of the dependent variable for the respective value of the independent variable. Centered on the regression line depicts a mean of zero for the random error. Moreover, the similarity in the spread of the bell curves denotes equal variance of the error term. In any specific instance, the random error term ε could be positive, negative, or zero, and the corresponding value of the dependent variable would be above, below, or on the regression line.

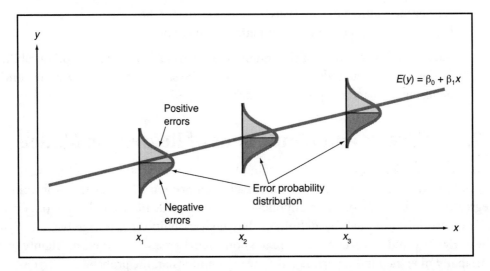

Figure 12.4   Decomposition of total variance into explained and unexplained variation

McClave, James T.; Benson, P. George; Sincich, Terry L.; *A First Course in Business Statistics*, 8th Ed., ©2001, p. 74. Reprinted and electronically reproduced by permission of Pearson Education, Inc., Upper Saddle River, New Jersey.

One of the common violations of these assumptions is that of equal variance of the error terms. As can be seen by comparing the two scatter diagrams in Figure 12.5, the observations are scattered uniformly about the regression line in Panel A but are scattered further from the regression line for higher values of X in Panel B. That is, if we were to plot the residuals or the error terms with respect to the independent variable X, for Panel A we would plot a uniform distribution as shown in Panel C. However, a similar plot for data in Panel B would yield an upward sloping graph as shown in Panel D. This is known as heteroscedasticity, or unequal variances of the residual term. A possible solution to the problem posed by heteroscedasticity is to partition the data and fit a regression model for each partition. This ensures that the variances are equal for each partition.

Predictions of the values of the dependent variable are meaningful only for the range for which data is available to determine the regression line. Prediction of dependent variable for the values of independent variable outside the range could result in nonsensical and counter-intuitive predictions. Consider an airline company using regression to determine the price to be charged for a short domestic flight of 150 miles, say, between Albany, New York, and New York City. Suppose it uses data from international flights with distances ranging from 1,000 miles to 10,000 miles. Its regression line could predict a negative fare for flights between Albany and New York City if the regression line has a large negative intercept. This is a meaningless result but is not due to the error in the method of regression but due to the limitation of data and erroneous usage to predict a value that was not in the range of data used to define the regression line.

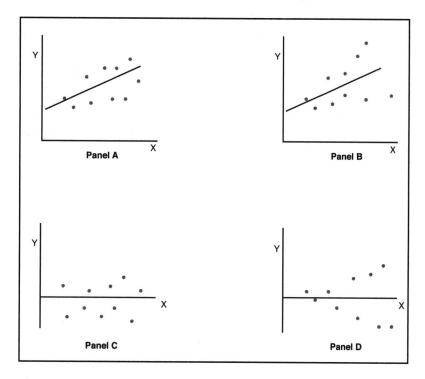

Figure 12.5   Regression model for second degree polynomial

It is important to note that the correlation coefficient or the coefficient of determination measures the degree of association between two variables. The association does not however indicate cause-and-effect relationship. One of the variables need not be identified as a cause and the other as an effect. Often humorous examples of high correlations have been given to highlight that correlation does not imply causation. Similarly, high house prices are usually correlated with high property taxes is misconstrued to imply that high property taxes cause high house prices. A high degree of correlation between two variables could be attributed to a third variable (or possibly others).

As regression and correlation are computed on sample data, the results are subject to sampling error. When interpreting or relying on the results of regression or correlation, the impact of sampling error has to be assessed. In a forensic accounting context, when presenting or being presented with statistical results from regression or correlation, focusing only on the parameters is insufficient. The forensic accountant must take the sampling error into account and question the confidence in the statistical results, in other words the credibility of statistical evidence.

The test of significance of the regression model relies on both the coefficient as well as the sample size. Hence, at times even for a low level of the coefficient of regression, the model may be deemed significant when the sample is very large. That is, if the sample size is 10,000 a very small coefficient R^2 such as 0.05 can be deemed significant. The inference is that there is a relationship between the two variables, even though only 5% of the variance of the dependent variable is accounted for by the variance in the independent variable. In such situations, the regression model, though statistically significant, might not be a reliable tool for forecasting values of the dependent variable based on observed value of the independent variable. This would be captured by a large standard error term, which will yield a wide prediction interval.

In interpreting results based on regression and correlation analysis, it is important to guard against spurious correlation. In spurious correlations there is no underlying theory or reason for the correlation, and often times the theory suggests otherwise. One such example is that life expectancy has doubled since the discovery of tobacco. It has been misconstrued to mean that consumption of tobacco increases longevity. In forensic accounting context, it is important to be cautious that the observed correlation is not spurious but can be corroborated through other evidence.

12.9 Other Regression Models

Only one form of the regression model, the ordinary least square regression, has been considered so far. Often in accounting and finance the relationships between variables are linear, and the theoretical considerations suggest that the linear regression is the form of model required. In general however, the determination of the most appropriate regression model should be based on a combination of theoretical construct, availability of data, and practical considerations. In some situations, the linear regression model may not be the best description of the underlying data. Sometimes the data on both variables might have to be transformed to deduce the relationship between the variables. At other times, different regression models may be deemed more applicable. This section provides a brief description of a few other regression models.

Logarithmic Regression Model

Sometimes when the dependent variable has wide variance and is very large relative to the independent variable, a logarithmic transformation may be desirable. In such instances, log Y is substituted in the linear regression equation. The regression equation then transforms to

$$\log y = \alpha + \beta x + \varepsilon$$

and may result in a better fit of the data. The variable Y no longer has to be normally distributed; instead, its logarithm is assumed to be normally distributed. Additionally, the interpretation of the slope of the regression line β changes. It is no longer to be interpreted as a constant amount of change per unit of change in X, rather a constant rate of change.

Logarithmic regression models could be specified by fitting an equation with logarithms of both variables as

$$\log y = \alpha + \beta \log x + \varepsilon.$$

The interpretation of the slope β is different. It is the rate of percentage change in the dependent variable per unit of percentage change in the independent variable. This formulation is popular in micro-economics in estimating the rate of change in consumption based on the rate of change in income. Keynesian economists believe that the marginal propensity to consume decreases as income increases and would result in the slope β being negative.

Polynomial Regression Models

Polynomial regression models are appropriate to represent the curvilinear relationship between two variables. Because a straight line is a polynomial of the first degree, a shape such as a parabola is a polynomial of the second degree. The resultant regression equation is of the form

$$y = \alpha + \beta_1 x + \beta_2 x^2 + \varepsilon.$$

Figure 12.6 shows two curvilinear relationships that can be modeled by the second degree polynomial regression model. Other shapes, such as the inverse or mirror image of the two shown in the figure, could also be modeled by reversing the signs on the slope coefficients. As can be seen in the figure or by analyzing the equation, a second degree polynomial allows for one change in direction; whereas, in the straight line equation, no change of direction is possible. In this equation, when β_1 and β_2 are of opposite signs, they would cancel each other at a particular value of x, and the value of y would be normally distributed around the mean equal to the intercept α. At that point the direction of the curve will change. Similarly, two changes in direction can be modeled through a third-degree polynomial.

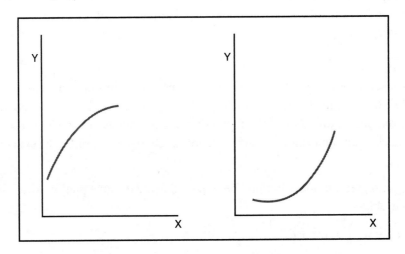

Figure 12.6   Curvilinear relationships

Multiple Regression Models

In certain situations, instead of a single independent variable, two or more independent variables are thought to have impacted the value of the dependent variable. In these cases, the regression equation is solved to estimate the marginal impact of each of the independent variables on the dependent variable. The models thus constructed could be linear, logarithmic, or polynomial. The R^2, or the coefficient of determination, is now a measure of the amount of variance in the dependent variable being explained by the variance in two or more independent variables.

In our example of housing prices, in the simple regression model the independent variable was the size of the house. In a multiple regression model, additional independent variables can be introduced, such as the lot size, the design of the house, the number of bathrooms, and so on. The model corresponding to a multiple regression with two independent variables is of the form

$$y = \alpha + \beta_1 x_1 + \beta_2 x_2 + \varepsilon$$

where x_1 and x_2 are the two independent variables. In general, the multiple regression model can include any number of independent variables. The regression coefficients β_1 measures the change in the dependent variable per unit of change in the independent variable x_1, when x_2 is fixed. Similarly, coefficients β_2 measures the change in the dependent variable per unit of change in the independent variable x_2, when x_1 is held fixed. The regression coefficients, β_1 and β_2, could have the same sign or opposite signs.

12.10 Chapter Summary

This chapter discussed statistical techniques of relating two continuous variables. Earlier you were introduced to terminology of relationship between two discrete variables in terms of conditional and joint probabilities. However, when one or both variables are continuous, enumerating the conditional probabilities is tedious; instead, the statistical methodology of regression and correlation is employed. The scatter plots or diagrams help in visualization of data and reveal whether or not there is a relationship between the variables. The relationship between two variables can be formally ascertained through the tools of correlation analysis and the strength of the relationship measured through a correlation coefficient.

The regression methodology is initially developed as a linear model between two variables. The "best-fitting" line is defined as one that minimizes the sum of squared errors when errors are computed as the vertical distance from the data points to the corresponding point on the straight line. The regression line is defined by two parameters, the intercept α, and the slope β. The regression line can be used to make estimates of point values by substituting the value of the independent variable and computing the value of the dependent variable.

Statistical metrics have been developed to evaluate the efficacy of the regression model. In addition to the test of significance for the individual parameters, the coefficient of determination measures the percentage of total variation in the dependent variable that has been accounted for by the regression line. The range of the value of the coefficient of determination, R^2, is between zero and one.

Assumptions and limitations of regression analysis were discussed in detail. Prior to interpreting the results of a regression model, it is important to evaluate the validity of the underlying assumptions of the model. Additionally, because the model generates a mathematical relationship that best fits the sample data, care has to be exercised in interpreting as well as relying on the model predictions. The mathematically determined relationship should not be overinterpreted. For example, correlation between two variables should not be interpreted as a proof of causation. Additionally, predictions using a regression model are valid only within the range of data used to develop the model. It might not be prudent to extrapolate the regression model well beyond the values of the data points used to construct the model.

Appendix: Use of Scientific Calculator and Excel

OLS Regression Equation:

Using Graphing Calculator such as TI-84

1. Press STAT 1
2. Enter data of dependent variable on L2 and that of independent variable on L1.
3. Press STAT and highlight CALC.
4. Press 4 for LinReg(ax + b), enter (L1, L2), and press Enter to compute.

The output gives the value of "b" or the intercept and the value of "a" or slope.

It also gives the value of "r" or correlation coefficient and the value of "r^2" the coefficient of determination.

If the values of "r" and "r^2" do not appear, follow these steps:

1. Turn the diagnostic feature on.
2. Press 2nd 0 for CATALOG.
3. Press "ALPHA" then the x^{-1} key for D.
4. Scroll down, highlight "DiagnosticsOn," and press Enter.

The values of "r" and "r^2" will now appear below the regression line. The scientific calculator also allows plotting the least square line.

1. Enter data on dependent and independent variables as before.
2. Press STAT, scroll right once, and highlight CALC.
3. Press 4 for LinReg(ax + b), enter (L1, L2, Y1), and then Press 2nd Y= for STATPLOT
4. Y1 can be found by pressing VARS, scrolling to the right to Y-VARS, and pressing "ENTER" twice.

The regression line is now plotted.

Using Excel:

Check in the "Data" toolbar for the "Data Analysis" option on the far right corner. If the option is not available, it has to be activated. The steps to activation are as follows:

1. Click on the Office Button at the top left corner and choose "Excel Options."
2. Click on "Add-Ins" on the left menu bar.

3. Locate "Analysis ToolPak-VBA" and activate it by selecting it and hitting "Go" at the bottom of the window.

4. Allow it to install, if needed.

Once the "Data Analysis" option is installed, you are able to access it by clicking on "Data" and then finding it on the left corner of the menu ribbon. The steps are as follows:

1. Enter the dependent variables in one column and independent variables in another.

2. Click on the "Data Analysis" command, a pop-up box opens.

3. Scroll down to "Regression" and press "OK".

4. Another pop-up window opens.

5. Select the cells with dependent variables on the "Input Y Range".

6. Select the cells with independent variables on the "Input X Range".

7. Choose a "Confidence Level" as desired. 95% is the default.

8. Choose preferred "Output options".

9. Check boxes for "Residuals", "Residual Plots", "Standardized Residuals" or "Line Fit Plots".

10. Click "OK". The regression results of slope, intercept, R-squared, and so on are displayed.

Index

A

a-priori probability, 111
acceptance sampling, 191
accounting rules, internal controls to ensure adherence to, 42
ACFE (Association of Certified Fraud Examiners), 11
Adelphia, 25
ADRs (American Depository Receipts), 36
adverse selection, 31
AEP Energy Services, 25
agency theory, 31
AICPA (American Institute of Certified Public Accountants), 27-28, 67
Albrecht, Steve, xvi
alternative hypothesis, 237
Ambani, Anil, 37
Ambani, Dhirubhai, 37
Ambani, Mukesh, 37
anticipated error rates, 226-228
application controls, 43
AS 5 (Auditing Standard No. 5), 67
assets
    internal controls for safeguarding of organizational, 41
    misappropriation of, 80
association analysis, 93-95

association rules, 94
AU (Audit Sampling) 350, 188
audit committee
    financial reporting process, role and responsibilities in, 70-75
    international guidelines, 72-73
    oversight function, 74-75
    overview, 34
    tips for, 75
auditors
    external auditors, role and responsibilities in financial reporting process of, 66-68
    hypergeometric distribution, use of, 145
    internal auditors, role and responsibilities in financial reporting process of, 78-80
audits
    expectations gap in, 65-66
    for public companies, history of requirement for, 61-62
    sampling
        acceptance, 191
        discovery, 190
        DUS (dollar unit sampling), 191
        overview, 188-190
        PPS (probability proportional to size), 191

B

back-up procedures, 44
balance sheet fraud, 8-9
batch control totals, 43
Bayes' rule, 116-122
belief system, 40
Bingham, James, 83
binomial distribution
    Excel used for statistical computations, 147
    mean (expected value), 130-131
    normal distribution, expressed as, 171
    overview, 128-132
    standard deviation, 131-132
    symmetrical, 132
Blue Ribbon Panel, 34
BOD (Board of Directors)
    audit committee, 34
    ethics of, 70
    financial reporting process, role and responsibilities in, 68-71
    oversight function, 68-69
    overview, 32-34
Boolean vectors, 94
boundary system, 40
Bush, George W., 24

C

Cadbury Report, 33
Cendant, 25
census, 180-181
Center of Audit Quality, 65, 77
Central Limit Theorem, 158, 199-200
CEO (Chief Executive Officer)
    BOD, influence on, 34
    perquisites and related party transactions, 56
certifications for forensic accountants, 26
Certified Public Accountant (CPA), 26
CFE (Certified Fraud Examiner), 26
CFF (Certified in Financial Forensics), 26
Chebycheff's inequality, 169
CIA (Certified Internal Auditor), 26
Citigroup, 2
cluster analysis, 96-97
cluster sampling, 185
coefficient of determination, 236, 241
combinatorial probability, 141
complement of event, 111
Comverse, 25
conditional probability, 111-117. *See also* Bayes' rule
confidence intervals
    for large sample when population standard deviation is known, 205-209
    for large sample when population standard deviation is unknown, 209-211
    overview, 203-205, 238-239
    for proportions, 212-214
    scientific calculator used to compute, 217
    for small samples, 211-212

steps for constructing, 215-216
tolerable error, 220-221
confidence levels and determining sample size, 223, 226-228
confidentiality for whistleblowers, 82
confirming evidence, 108-109
consistent estimators, 202
continuous probability distributions
  binomial distribution expressed as a normal distribution, 171
  Chebycheff's inequality, 169
  cumulative distribution function, 150-156
  exponential distribution, 172-173
  joint distribution of continuous random variables, 173-176
  normal probability distribution, 157-170
  overview, 149-156
  probability distribution function, 150-153
  uniform (rectangular) probability distribution, 151-153, 156-157
control activities, 47, 50
control environment, 47, 49-50
Cooper, Cynthia, 78
corporate fraud
  earliest instance of, 61-62
  FBI's role in detecting, 26
  investigations, 5
  overview, 5
  whistleblowers, groups who have been, 81

corporate governance
  agency theory, 31
  BOD (Board of Directors), 32-34
  in developed economies, 33-34
  in emerging economies, 35-39
  organizational controls, 39-41
  overview, 31-33
corrective controls, 42
correlation
  overview, 233-234
  sampling error, 242
correlation coefficient, 234
corruption, 80
COSO (Treadway Commission's Committee of Sponsoring Organization) framework on internal controls, 45-52, 57-58
CPA (Certified Public Accountant), 26
Crazy Eddie, 17
Cressey, Donald, xvi
cumulative distribution function, 150-156
curvilinear relationships, 233, 243
cyber-crime, 16-18
cyber-security, use of Poisson distribution in, 140

D

data classification, 92
data encryption, 44
data mining
  association analysis, 93-95
  cluster analysis, 96-97

data classification, 92
money laundering, detection of, 100-103
outlier analysis, 98-100
overview, 89-91
deficiency in internal controls, 51
Deloitte Forensic Center, 75
detective controls, 42
determining sample size
anticipated error rates, for differing, 226-228
confidence, for differing levels of, 223, 226-228
overview, 219
precision, for differing levels of, 223, 226-228
for proportions, 225-228
when population deviation is known, 220-221
when population deviation is unknown, 222-225
developed economies, corporate governance in, 33-34
diagnostic control system, 41
discovery sampling, 190
discrete probability distributions
binomial distribution
Excel used for statistical computations, 147
mean (expected value), 130-131
overview, 128-132
standard deviation, 131-132
symmetrical, 132
hypergeometric distribution
auditors' use of, 145
Excel used for statistical computations, 147
mean (expected value), 142
multilocation inventory audit example, 142-145
overview, 140-142
standard deviation, 142
mean (expected value), 126-127
overview, 126
Poisson distribution
application of, 137
cyber-security, use in, 140
Excel used for statistical computations, 147
mean (expected value), 137
overview, 132-136
standard deviation, 137
variance, 137
standard deviation, 127
document matching, 44
Dodd-Frank Wall Street Reform and Consumer Protection Act, 23, 33, 80
DOJ (Department of Justice) and prosecution of criminals committing fraud, 24-25
DUS (dollar unit sampling), 191
Dyck, A., 81
Dynegy, 25

E

earnings management, 77
edit checks, 44
efficiency, internal controls promoting, 42
efficient estimators, 202
embezzlement, 16, 54

emerging economies, corporate governance in, 35-39
employee collusion, 94-95
employee detection of management fraud, 82
employee fraud
    embezzlement, 16, 54
    fraudulent write-offs, 13
    ghost employees, 14-15
    inventory shrinkage, 15-16
    lapping, 13
    overview, 11
    shell companies, 13-14
    skimming, 12
Enron, 14, 24, 188-189
Enterasys, 24
errors
    anticipated error rates, 226-228
    random error, 232-233, 239-241
    sampling error, 197, 242
    standard error, 236
    tolerable error, 220-221
    Type I error, 238
    Type II error, 238
estimation of population parameters, 200-203
ethical culture, 27-28
ethics of BOD, 70
events, 111
evidence
    confirming, 108-109
    information and, 110
    probative value of, 113-117
    schematic representation of, 108-109
    smoking gun, 188-189

Excel used for statistical computations
    binomial distribution, 146
    hypergeometric distribution, 146-147
    Poisson distribution, 147
expectations gap in accounting profession, 65-66
exponential distribution, 172-173
external audit of internal controls, 45-46
external auditors, role and responsibilities in financial reporting process of, 66-68

F

family relationships in emerging economies, 35-39
FCPA (Foreign Corrupt Practices Act), 75
Federal Civil False Claims Act, 82
financial firm failures, 63
financial institution fraud, 6
Financial Reporting Council of the U.K., 72
financial reporting process as collective effort
    audit committee, 70-75
    Board of Directors, 68-71
    external auditors, 66-68
    internal auditors, 78-80
    management, 75-77
    overview, 63-66
    whistleblower, role of, 80-84
firewalls, 44
Ford, Joseph, 16

forensic accountants
  certifications, 26
  qualifications required for, 26
  responsibilities of, 26
Forensic Analytics (**Nigrini**), xvi
Fornelli, Cynthia, 65
fraud
  characteristics of, 3
  corporate
    earliest instance of, 61-62
    FBI's role in detecting, 26
    investigations, 5
    overview, 5
    whistleblowers, groups who have been, 81
  cyber-crime, 16-18
  DOJ prosecution of criminals committing, 24-25
  earliest instance of, 61-62
  employee
    embezzlement by management, 16
    fraudulent write-offs, 13
    ghost employees, 14-15
    inventory shrinkage, 15-16
    lapping, 13
    overview, 11
    shell companies, 13-14
    skimming, 12
  financial institution, 6
  health care, 6
  investigations of, 4-5
  management
    balance sheet fraud, 8-9
    employee detection of, 82
    HealthSouth, 9-10
    inflating revenue, schemes for, 7
    Lehman, 10-11
    motivation and opportunity for, 77
    overview, 7-9, 11
    understating expenses, schemes for, 7-8
    WorldCom, 9
  mortgage, 6
  overview, 3-7
  securities, 5
  types of, 3
Fraud Triangle (**Cressey**), xvi
fraudulent write-offs, 13

G

Galleon Management, 71
Gaussian probability distribution. *See* **normal probability distribution**
general internal controls, 43
ghost employees, 14-15
Goldman Sachs, 71, 109
Gosset, W.S., 212
Groupon, 45-46

H

health care fraud, 6
HealthSouth, 9-10
Homestore, 25
HSBC, 100
hypergeometric distribution
  auditors' use of, 145
  Excel used for statistical computations, 147

mean (expected value), 142
multilocation inventory audit example, 142-145
overview, 140-142
standard deviation, 142

I

ICFRs (internal controls over financial reporting), 74
inflating revenue, schemes for, 7
information
    and communication, 48, 50
    and evidence, 110
infoUSA Inc., 56
inside whistleblowers, 83-82
insider trading, 71, 109
Institute of Internal Auditors, 78
interactive control system, 41
internal auditors, roles and responsibilities in financial reporting process, 78-80
internal controls
    application, 43
    benefits of, 52
    common, 43-44
    corrective, 42
    COSO framework on, 45-52
    costs of, 52-53, 58
    deficiency in, 51
    detective, 42
    external audit of, 45-46
    general, 43
    legislation on, 58
    limitations of, 53-56
    management override, 55-56

outsourcing and, 51
overview, 41-42
preventive, 42
international guidelines for audit committee, 72-73
interval estimate, 203. *See also* confidence intervals
inventory shrinkage, 15-16
investigations of fraud, 4-5

J

joint distribution of continuous random variables, 173-176
joint event, 111
joint probability, 106, 112-113, 115
judgment (nonrandom) sampling, 181-182, 187-189

K

Ka-Shing, Li, 38
Koss Corporation, 54
Kreuger and Toll, 62

L

lapping, 13
large sample
    confidence intervals when population standard deviation is known, 205-209
    confidence intervals when population standard deviation is unknown, 209-211

Lee, Matthew, 83
legislation
    Dodd-Frank Wall Street Reform and Consumer Protection Act, 23, 33, 80
    on internal controls, 58
    overview, 21
    SOX (Sarbanes-Oxley Act of 2002), 22-23, 33, 76
Lehman Brothers, 10-11, 83
Levers of Control (Simons), 39
Levitt, Arthur, 70
likelihood ratio, 116-117
linear relationship, 233
lockbox system, 44
logarithmic regression model, 243

M

Madoff, Bernie, 2
major nonconformity, 51
management
    financial reporting process, role and responsibilities in, 75-77
    internal controls, override of, 55-56
    policies, internal controls ensuring adherence to, 42
management fraud
    balance sheet fraud, 8-9
    employee detection of, 82
    HealthSouth, 9-10
    inflating revenue, schemes for, 7
    Lehman, 10-11
    motivation and opportunity for, 77
    overview, 7-9, 11
    understating expenses, schemes for, 7-8
    WorldCom, 9
marginal probability, 111, 115
mass marketing fraud, 7
material weakness, 51, 58
mathematical notations and formulae for probability, 111
McKesson and Robbins, 62
mean (expected value)
    binomial distribution, 130-131
    hypergeometric distribution, 142
    normal probability distribution, 158-159
    overview, 126-127
    Poisson distribution, 137
    uniform (rectangular) probability distribution, 157
Mercury Finance, 25
Merrill Lynch, 189
minor nonconformity, 51
money laundering
    detection of, 100-103
    overview, 100
    settlements in money laundering charges, recent, 100
monitoring activities, 48, 51
moral hazard, 33
Morse, A., 81
mortgage fraud, 6
multilocation inventory audit example for hypergeometric distribution, 142-145
multiple regression models, 244-245
mutual exclusivity, 106

N

Nigerian email/letter scheme, 6-7
Nigrini, Mark, xvi
nonparamethic outlier analysis, 99
nonrandom sampling, 181-182, 187-189
normal probability distribution
    binomial distribution expressed as, 171
    mean, 158-159
    overview, 157-170
    scientific calculator used to compute probabilities, 178
    standard deviation, 158-159
    standard normal distribution table, 162
    testing for normality, 168-170
normal probability plot, 169-170
null hypothesis, 237-238
numbered documents, 43
numeric value, 111

O

objective probability, 111
OLS (Ordinary Least Squares) regression, 234-235, 239-241
organizational controls, 39-41
organizational culture, 39
outlier analysis, 98-100
outsourcing and internal controls, 51
oversight function
    audit committee, 74-75
    BOD, 68-69

P

parametric outlier analysis, 99
PCAOB (Public Company Accounting Oversight Board), 22, 63, 67
perquisites and related party transactions, 56
physical security, 44
PNC, 25
point estimate, 203
Poisson distribution
    application of, 137
    cyber-security, use in, 140
    Excel used for statistical computations, 147
    mean (expected value), 137
    overview, 132-136
    standard deviation, 137
    variance, 137
polynomial regression models, 243-244
Ponzi, Charles, 1
Ponzi scheme, 1-2, 62
population deviation is known, determining sample size when, 220-221
population deviation is unknown, determining sample size when, 222-225
population parameters
    estimation of, 200-203
    generalizing sample data to, 196-199
    interval estimate, 203
    point estimate, 203
posterior probability, 121-122

PPS (probability proportional to size) sampling, 191
precision levels, determining sample size for differing, 223, 226-228
prediction intervals, 238-239
President's Corporate Fraud Task Force, 24
preventive controls, 42
principal-agent conflict, 35
principal-principal relationship, 36
probabilistic linear models, 232-233
probability
    a-priori, 111
    Bayes' rule, 116-122
    combinatorial, 141
    concepts and terminology, 106
    conditional, 111-115
    distributions, 125
    information and evidence, 110
    joint, 106, 112-113, 115
    likelihood ratio, 116-117
    marginal, 111, 115
    mathematical notations and formulae, 111
    mutual exclusivity, 106
    numeric value, 111
    objective, 111
    overview, 106
    posterior, 121-122
    probative value of evidence, 113-117
    sample space, 106
    schematic representation of evidence, 108-109
    subjective, 111
    Venn diagrams, 106-109
probability distribution function, 150-153
probative value of evidence, 113-117
Proctor and Gamble, 71
professional guidance for internal auditors, 78-79
proportions
    confidence intervals, 212-214
    determining sample size, 225-228
Public Company Accounting Oversight Board (PCAOB), 22, 63, 67

Q

qualifications required for forensic accountants, 26
Qwest, 25

R

random error, 232-233, 239-241
random sampling
    cluster, 185
    overview, 181-184
    selection with replacement, 182-183
    selection without replacement, 182-184
    stratified, 184-185
    systemic, 184

reconciliation of accounts, 43
recording of business transactions, internal controls for accuracy in, 41
Refco, 25
regression
    alternative hypothesis, 237
    coefficient of determination, 236, 241
    confidence intervals, 238-239
    logarithmic regression model, 243
    multiple regression models, 244-245
    null hypothesis, 237-238
    OLS regression, 234-235, 239-241
    overview, 234-236
    polynomial regression models, 243-244
    prediction intervals, 238-239
    prediction using, 238-239
    random error, 232-233, 239-241
    sampling error, 242
    test of significance, 237-238
    Type I error, 238
    Type II error, 238
regression line, 234-236
regulation
    audits for public companies, history of requirement for, 61-62
    DOJ, fraud prosecution by, 24-25
    internal controls for compliance with, 42
    overview, 21
Reliance Industries, 37
Repo 105, 10-11, 83

"Responsibility for Preventing and Detecting Financial Reporting Fraud" (SEC panel discussion), 65-66
revenue, schemes for inflating, 7
risk assessment
    with COSO framework, 57-58
    overview, 47, 50
risk-based approach to audit procedure, 67
Romania, cyber-crime in, 18
round trip transaction, 189

S

sample mean, 198-199
sample space, 106
sample standard deviation, 198-199
sampling
    audit
        acceptance, 191
        discovery, 190
        DUS (dollar unit sampling), 191
        overview, 188-190
        PPS (probability proportional to size), 191
    census compared, 180-181
    determining sample size
        anticipated error rates, for differing, 226-228
        confidence, for differing levels of, 223, 226-228
        overview, 219
        precision, for differing levels of, 223, 226-228

for proportions, 225-228
  when population deviation is known, 220-221
  when population deviation is unknown, 222-225
  judgment (nonrandom), 181-182, 187-189
  overview, 179-181
  random
    cluster, 185
    overview, 181-184
    selection with replacement, 182-183
    selection without replacement, 182-184
    stratified, 184-185
    systemic, 184
sampling error, 197, 242
SAS (Statement of Auditing Standards) No. 99, 27-28, 67
scatter plot, 233
schematic representation of evidence, 108-109
scientific calculator
  for confidence intervals, 217
  for normal probability distribution, 178
  for OLS regression, 247
SEC (Securities and Exchange Commission)
  overview, 5
  SOX, study and clarification of, 23
Securities and Exchange Act of 1933, 70

securities fraud, 5
segregation of duties, 43
settlements in money laundering charges, recent, 100
shareholders in emerging economies, controlling, 35-39
shell companies, 13-14
shopping cart analysis, 93
significant deficiency, 51
Simons, Robert, 39
simple event, 111
skimming, 12
small samples, confidence intervals for, 211-212
smoking gun evidence, 188-189
SOX (Sarbanes-Oxley Act of 2002), 22-23, 33, 76
SSE (sum of the squared errors), 234
Standard Chartered PLC, 100
standard deviation
  binomial distribution, 131-132
  hypergeometric distribution, 142
  normal probability distribution, 158-159
  overview, 127
  Poisson distribution, 137
  uniform (rectangular) probability distribution, 157
standard error, 236
standard normal distribution table, 162
Stanford International, 2

statistical inference from sample information
   Central Limit Theorem, 199-200
   confidence intervals
      for large sample when population standard deviation is known, 205-209
      for large sample when population standard deviation is unknown, 209-211
      overview, 203-205
      for proportions, 212-214
      scientific calculator used to compute, 217
      for small samples, 211-212
      steps for constructing, 215-216
   overview, 195-196
   population parameters
      estimation of, 200-203
      generalizing sample data to, 196-199
      interval estimate, 203
      point estimate, 203
   sample mean, 198-199
   sample standard deviation, 198-199
   sampling error, 197
stratified sampling, 184-185
Student t-distribution, 212
subjective probability, 111
sufficient estimator, 202
symmetrical binomial distribution, 132
system/process documentation, 43
systemic sampling, 184

T

t-distribution, 212
test of significance, 237-238
testing for normality, 168-170
tolerable error, 220-221
Treadway Commission's Committee of Sponsoring Organization (COSO) framework on internal controls, 45-52, 57-58
Tyco International, 16
Type I error, 238
Type II error, 238

U

unbiased estimators, 202
unbundling, 6
understating expenses, schemes for, 7-8
uniform (rectangular) probability distribution, 151-153, 156-157

V

Valukas, Anton, 10
Vanda Computers, 38
variance
   Poisson distribution, 137
   uniform (rectangular) probability distribution, 157
Venn diagrams, 106-109

W

Wells, Joseph, xvi
whistleblowers
   confidentiality for, 82
   costs to, 83-82
   financial reporting process, role and responsibilities in, 80-84
   groups who have been, 81
   inside, 83-82
"Who Blows the Whistle on Corporate Fraud?" (Dyck, Morse, and Zingales), 81
WorldCom, 9, 25, 78

X

Xerox, 83

Z

Zingales, L., 81

In an increasingly competitive world, it is quality of thinking that gives an edge—an idea that opens new doors, a technique that solves a problem, or an insight that simply helps make sense of it all.

We work with leading authors in the various arenas of business and finance to bring cutting-edge thinking and best-learning practices to a global market.

It is our goal to create world-class print publications and electronic products that give readers knowledge and understanding that can then be applied, whether studying or at work.

To find out more about our business products, you can visit us at www.ftpress.com.